SpringerBriefs in Electrical and Computer Engineering

Computational Electromagnetics

Series editor

Rakesh Mohan Jha, Bangalore, India

More information about this series at http://www.springer.com/series/13885

Hema Singh · R. Chandini
Rakesh Mohan Jha

EM Design and Analysis of Dipole Arrays on Non-planar Dielectric Substrate

 Springer

Hema Singh
Centre for Electromagnetics
CSIR-National Aerospace Laboratories
Bangalore, Karnataka
India

Rakesh Mohan Jha
Centre for Electromagnetics
CSIR-National Aerospace Laboratories
Bangalore, Karnataka
India

R. Chandini
Centre for Electromagnetics
CSIR-National Aerospace Laboratories
Bangalore, Karnataka
India

ISSN 2191-8112   ISSN 2191-8120   (electronic)
SpringerBriefs in Electrical and Computer Engineering
ISSN 2365-6239   ISSN 2365-6247   (electronic)
SpringerBriefs in Computational Electromagnetics
ISBN 978-981-287-780-2   ISBN 978-981-287-781-9   (eBook)
DOI 10.1007/978-981-287-781-9

Library of Congress Control Number: 2015947795

Springer Singapore Heidelberg New York Dordrecht London

Printed on acid-free paper

Springer Science+Business Media Singapore Pte Ltd. is part of Springer Science+Business Media
(www.springer.com)

To Dr. Sudhakar K. Rao

In Memory of Dr. Rakesh Mohan Jha
Great scientist, mentor, and excellent
human being

Dr. Rakesh Mohan Jha was a brilliant contributor to science, a wonderful human being, and a great mentor and friend to all of us associated with this book. With a heavy heart we mourn his sudden and untimely demise and dedicate this book to his memory.

Preface

Low profile antennae are of great interest for antenna designers and engineers toward conformity and low observability. This book presents the EM design and analysis of printed dipole array on planar and cylindrical substrate. The substrate is taken as low-loss dielectric. The effect of infinite and finite ground plane on antenna array performance is discussed. The performance of linear and planar dipole array is analyzed in terms of input impedance, return loss, and radiation pattern. This book is divided into sub-sections. First, the design and performance analysis of a single half-wave center-fed dipole over a planar and nonplanar substrate is presented. The performance analysis of antenna is carried out in terms of input impedance, return loss, and radiation pattern. Then the design and performance analysis of linear dipole arrays of different configurations is presented. Illustrations for both planar and cylindrical substrate are included. Next, the EM design and analysis of planar dipole arrays on a cylindrical substrate are discussed. The study presented is geared toward the EM design and analysis of conformal arrays. This book serves as a systematic step-by-step guide for beginners in low-profile antenna design.

Hema Singh
R. Chandini
Rakesh Mohan Jha

Acknowledgments

We would like to thank Mr. Shyam Chetty, Director, CSIR-National Aerospace Laboratories, Bangalore for his permission and support to write this SpringerBrief.

We would also like to acknowledge the valuable suggestions from our colleagues at the Centre for Electromagnetics, Dr. R.U. Nair, Dr. Shiv Narayan, Dr. Balamati Choudhury, and Mr. K.S. Venu during the course of writing this book. We express our sincere thanks to Mr. Harish S. Rawat, Ms. Neethu P.S., Mr. Umesh V. Sharma, and Mr. Bala Ankaiah, the project staff at the Centre for Electromagnetics, for their consistent support during the preparation of this book.

But for the concerted support and encouragement from Springer, especially the efforts of Suvira Srivastav, Associate Director, and Swati Meherishi, Senior Editor, Applied Sciences & Engineering, it would not have been possible to bring out this book within such a short span of time. We very much appreciate the continued support by Ms. Kamiya Khatter and Ms. Aparajita Singh of Springer toward bringing out this brief.

Contents

EM Design and Analysis of Dipole Arrays on Non-planar
Dielectric Substrate 1
1   Introduction 1
2   Single Dipole on Planar Ground Plane 3
   2.1   EM Design 3
   2.2   Performance Analysis 5
3   Dipole Array Design 17
   3.1   Two-Dipole Array Over Cylindrical Surface 18
   3.2   Three-Dipole Array 21
   3.3   Five-Dipole Array 33
   3.4   Ten-Dipole Array on Cylindrical Substrate 36
4   Planar Dipole Array 38
   4.1   A Square Dipole Array Over a Cylindrical Surface 38
   4.2   A Rectangular Dipole Array Over a Cylindrical Surface 45
5   Conclusion 67
References 67

About the Book 69

Author Index 71

Subject Index 73

About the Authors

Dr. Hema Singh is currently working as Senior Scientist in Centre for Electromagnetics of CSIR-National Aerospace Laboratories, Bangalore, India. Earlier, she was Lecturer in EEE, BITS, Pilani, India during 2001–2004. She obtained her Ph.D. degree in Electronics Engineering from IIT-BHU, Varanasi India in 2000. Her active area of research is Computational Electromagnetics for Aerospace Applications. More specifically, the topics she has contributed to, are GTD/UTD, EM analysis of propagation in an indoor environment, phased arrays, conformal antennas, radar cross section (RCS) studies including Active RCS Reduction. She received Best Woman Scientist Award in CSIR-NAL, Bangalore for period of 2007–2008 for her contribution in the areas of phased antenna array, adaptive arrays, and active RCS reduction. Dr. Singh has co-authored one book, one book chapter, and over 120 scientific research papers and technical reports.

R. Chandini obtained her BE (ECE) degree from Visvesvaraya Technological University, Karnataka. She was a Project Engineer at the Centre for Electromagnetics of CSIR-National Aerospace Laboratories, Bangalore, where she worked on RCS studies and conformal arrays.

Dr. Rakesh Mohan Jha was Chief Scientist & Head, Centre for Electromagnetics, CSIR-National Aerospace Laboratories, Bangalore. Dr. Jha obtained a dual degree in BE (Hons.) EEE and M.Sc. (Hons.) Physics from BITS, Pilani (Raj.), India, in 1982. He obtained his Ph.D. (Engg.) degree from Department of Aerospace Engineering of Indian Institute of Science, Bangalore in 1989, in the area of computational electromagnetics for aerospace applications. Dr. Jha was a SERC (UK) Visiting Post-Doctoral Research Fellow at University of Oxford, Department of Engineering Science in 1991. He worked as an Alexander von Humboldt Fellow at the Institute for High-Frequency Techniques and Electronics of the University of Karlsruhe, Germany (1992–1993, 1997). He was awarded the Sir C.V. Raman Award for Aerospace Engineering for the Year 1999. Dr. Jha was elected Fellow of INAE in 2010, for his contributions to the EM Applications to Aerospace

Engineering. He was also the Fellow of IETE and Distinguished Fellow of ICCES. Dr. Jha has authored or co-authored several books, and more than five hundred scientific research papers and technical reports. He passed away during the production of this book of a cardiac arrest.

List of Figures

Figure 1   Cartesian coordinate system used for EM design
            of a dipole.   4
Figure 2   Substrate for dipole design   4
Figure 3   Dipole arms and the feed gap on the substrate   5
Figure 4   Lumped port excitation at the feed gap of dipole   6
Figure 5   Ground plane below the dipole and substrate   7
Figure 6   Radiation boundary covering the dipole
            and the substrate   8
Figure 7   Port field display at dipole excitation   9
Figure 8   E-field distribution over the dipole antenna   9
Figure 9   Surface current over the excited dipole   10
Figure 10   Radiation pattern ($\phi = 0°$; $\phi = 90°$) of a single dipole
            antenna over an infinite rectangular ground plane.
            a Rectangular plot. **b** Polar plot   11
Figure 11   Impedance of a single dipole antenna over an infinite
            ground plane   12
Figure 12   Return loss of a single dipole antenna over an infinite
            ground plane   12
Figure 13   3-D radiation pattern of a single dipole antenna over an
            infinite ground plane. **a** Total gain. **b** Gain (ϕ).
            c Gain (θ)   13
Figure 14   Radiation pattern of a single dipole antenna over a finite
            rectangular ground plane. **a** Rectangular plot. **b** Polar
            plot   14
Figure 15   Impedance (real and imaginary) plot of a single dipole
            antenna over a finite ground plane   15
Figure 16   Return loss of a single dipole antenna over a finite
            ground plane   15
Figure 17   3-D radiation pattern of a single dipole antenna over
            a finite ground plane. **a** Total gain. **b** Gain (ϕ).
            c Gain (θ)   16

Figure 18   A single dipole over a cylinder 16
Figure 19   Lumped port excitation at the feed gap
            of printed dipole 17
Figure 20   Port field display at the feed gap of printed dipole 17
Figure 21   E-field distribution over the printed dipole. 17
Figure 22   Surface current over the printed dipole 18
Figure 23   Impedance (real and imaginary) of a single
            dipole antenna over a cylindrical surface................ 18
Figure 24   Return loss of a single dipole antenna over a cylindrical
            surface 19
Figure 25   Radiation pattern of a single dipole antenna over
            a cylindrical substrate. **a** Rectangular plot. **b** Polar
            plot 20
Figure 26   3-D radiation pattern of a single dipole antenna over a
            cylindrical surface. **a** ϕ versus θ. **b** θ versus ϕ 21
Figure 27   A single dipole over a cylindrical surface
            with cylindrical ground plane....................... 21
Figure 28   Performance of a single dipole antenna over a cylindrical
            surface with cylindrical ground plane. **a** Impedance plot.
            b Return loss.................................... 22
Figure 29   3-D radiation pattern of a single dipole antenna over a
            cylindrical surface with cylindrical ground plane.
            a Rectangular. **b** Polar. **c** 3-D 23
Figure 30   A single dipole over a cylindrical surface with truncated
            ground plane 24
Figure 31   Performance of a single dipole antenna over a cylindrical
            surface with truncated ground plane. **a** Impedance plot.
            b Return loss.................................... 25
Figure 32   3-D radiation pattern of a single dipole antenna over a
            cylindrical surface. **a** Rectangular. **b** Polar. **c** 3-D.......... 26
Figure 33   Single dipole antenna over a cylindrical surface with
            truncated ground plane with reduced height. **a** Model.
            b Return loss. **c** and **d** Radiation pattern 28
Figure 34   Single dipole antenna over a cylindrical surface
            with sector (120°) ground plane. **a** Model. **b** Return loss.
            c Radiation pattern 30
Figure 35   Two-dipole array over a cylindrical surface with trun-
            cated ground plane............................... 32
Figure 36   Lumped port excitation of two-dipole array over
            a cylindrical surface with truncated ground plane.
            a Dipole 1. **b** Dipole 2 32
Figure 37   Field port display of two-dipole array over a cylindrical
            surface with truncated ground plane. **a** Dipole 1.
            b Dipole 2...................................... 33

Figure 38 E-field distribution of two-dipole array over a cylindrical
           surface with truncated ground plane. **a** Dipole 1.
           b Dipole 2.   34
Figure 39 Surface current distribution of two-dipole array over
           a cylindrical surface with truncated ground plane.
           a Dipole 1. **b** Dipole 2   35
Figure 40 Impedance (real and imaginary) plot of two-dipole array
           over a cylindrical surface with truncated
           ground plane   36
Figure 41 Return loss of two-dipole array over a cylindrical surface
           with truncated ground plane.   37
Figure 42 Radiation pattern of a two-dipole antenna over
           a cylindrical substrate with truncated ground plane.
           a Rectangular plot. **b** Polar plot. **c** 3-D   38
Figure 43 Three-dipole array on a planar substrate. **a** Side view.
           b Top view   40
Figure 44 E-field distribution over three-dipole array antenna   41
Figure 45 Radiation pattern of a three-dipole array over a planar
           substrate and an infinite ground plane. **a** Rectangular
           plot. **b** Polar plot. **c** 3-D plot   41
Figure 46 Radiation pattern of three-dipole array antenna over
           a planar substrate and finite ground plane. **a** Rectangular
           plot. **b** Polar plot. **c** 3-D plot   43
Figure 47 Three-dipole array over a cylindrical surface
           with truncated ground plane.   44
Figure 48 Impedance (real and imaginary) plot of a three-dipole
           array over a cylindrical surface with truncated ground
           plane   45
Figure 49 Return loss of a three-dipole array over a cylindrical
           surface with truncated ground plane   46
Figure 50 Radiation pattern of a three-dipole array over
           a cylindrical surface with truncated ground plane.
           a Rectangular plot. **b** Polar plot. **c** 3-D plot   46
Figure 51 Five-dipole array over a planar substrate. **a** Side view.
           b Top view   48
Figure 52 E-field distribution over a five-dipole array on a planar
           substrate   49
Figure 53 Radiation pattern of a five-dipole array antenna
           over a planar substrate and an infinite ground plane.
           a Rectangular plot. **b** Polar plot. **c** 3-D plot   49
Figure 54 Radiation pattern of a five-dipole array antenna on planar
           substrate and finite ground plane. **a** Rectangular plot.
           b Polar plot. **c** 3-D plot.   51

Figure 55   Five-dipole array over a cylindrical surface
             with truncated ground plane.   52
Figure 56   Impedance (real and imaginary) plot of a five-dipole
             array over a cylindrical surface with truncated ground
             plane   53
Figure 57   Return loss of a five-dipole array over a cylindrical
             surface with truncated ground plane   54
Figure 58   Radiation pattern of a five-dipole array over a cylindrical
             surface with truncated ground plane. **a** Rectangular plot.
             b Polar plot. **c** 3-D plot.   54
Figure 59   Ten-dipole array over a cylindrical surface with ground
             plane. **a** Front view. **b** Back view. **c** Zoomed view.   56
Figure 60   Impedance (real and imaginary) plot of a ten-dipole array
             over a cylindrical surface with ground plane   58
Figure 61   Return loss of a ten-dipole array over a cylindrical
             surface with ground plane   58
Figure 62   Radiation pattern of a ten-dipole array over a cylindrical
             surface with ground plane   59
Figure 63   3 × 3 dipole array over a cylindrical surface with ground
             plane   60
Figure 64   Impedance (real and imaginary) plot of a 3 × 3 dipole
             array over a cylindrical surface with ground plane   61
Figure 65   Return loss of a 3 × 3 dipole array over a cylindrical
             surface with ground plane   62
Figure 66   Radiation pattern of a 3 × 3 dipole array over
             a cylindrical surface with ground plane. **a** Rectangular.
             b Polar. **c** 3-D   62
Figure 67   2 × 5 dipole array over a cylindrical surface with ground
             plane   64
Figure 68   Impedance (real and imaginary) plot 2 × 5 dipole array
             over a cylindrical surface with ground plane   65
Figure 69   Return loss of a 2 × 5 dipole array over a cylindrical
             surface with ground plane   65
Figure 70   Radiation pattern of a 2 × 5 dipole array over
             a cylindrical surface with ground plane. **a** Rectangular.
             b Polar. **c** 3-D   66

List of Tables

Table 1   Positions of two dipoles on the cylindrical substrate
with truncated ground plane   33
Table 2   Positions of three dipoles on the cylindrical substrate.   45
Table 3   Positions of five dipoles on the cylindrical substrate   53
Table 4   Positions of ten dipoles on the cylindrical substrate   57
Table 5   Positions of 3×3 dipoles on the cylindrical substrate   61
Table 6   Positions of 2×5 dipoles on the cylindrical substrate   64

EM Design and Analysis of Dipole Arrays on Non-planar Dielectric Substrate

Abstract This book presents a simple and systematic description of EM design of antenna arrays. Printed dipole antennas are known to be simple yet more efficient than wire antennas. The dielectric substrate and the presence of ground plane affect the antenna performance and the resonant frequency is shifted. This book includes the EM design and performance analysis of printed dipole arrays on planar and cylindrical substrates. The antenna element is taken as half-wave center-fed dipole. The substrate is taken as low-loss dielectric. The effect of substrate material, ground plane, and the curvature effect is discussed. Results are presented for both the linear and planar dipole arrays. The performance of dipole array is analyzed in terms of input impedance, return loss, and radiation pattern for different configurations. The effect of curved platform (substrate and ground plane) on the radiation behavior of dipole array is analyzed. The book explains the fundamentals of EM design and analysis of dipole antenna array through numerous illustrations. It is essentially a step-by-step guide for beginners in the field of antenna array design and engineering.

Keywords Printed dipole array · Planar and non-planar substrate · EM design · Radiation pattern

1 Introduction

This book presents the design of dipole array over planar and non-planar substrates. The dipole element is taken as a strip dipole in which the dipole arms are rectangular strips of width w and thickness t, where $w \gg t$. The strip dipoles can be easily fabricated on dielectric substrate and are commonly used in linear and planar arrays at microwave frequencies. Moreover, the strip dipoles in free space are complementary to the slots cut in thin ground planes as per Babinet's principle (Elliot 1981). Every antenna has advantages and disadvantages; printed dipole antennas have the disadvantage of large geometry. However, its size gets reduced at higher

© The Author(s) 2016
H. Singh et al., *EM Design and Analysis of Dipole Arrays on Non-planar Dielectric Substrate*, SpringerBriefs in Computational Electromagnetics,
DOI 10.1007/978-981-287-781-9_1

frequencies. The general principle, that thicker dipole provides wider bandwidth (Kim et al. 2007), is also applicable to a printed strip dipole antenna. The printed dipole antennas structure can be planar or curved (He and Wang 2007). These antennas are simple to design but more efficient than wire dipole antennas (Kim et al. 2007).

The antenna elements are printed on the dielectric substrate, which in turn affects its performance. The resonant frequency gets shifted due to the constitutive parameters of the substrate (Wium 2013). This makes the dielectric substrate material an important parameter in the antenna design. The selection of dielectric material of the substrate depends on the application and its desired radiation characteristics. In dielectric materials with higher permittivity, the surface wave excitations are higher but the bandwidth will be less (Kraus et al. 2006). If the thickness of the substrate is increased, wider bandwidth can be achieved. However, it has adverse effect of enhanced surface wave excitation. In other words, there is a trade-off between bandwidth and surface wave excitation for a given substrate. Moreover, the substrate material characteristics (permittivity and loss tangent) are frequency dependent, affecting in turn the efficiency of the antenna array.

The presence of ground plane in the antenna design is another important factor in determining the antenna performance. Moreover, the effect of ground plane is different for planar and curved surfaces (Yang et al. 2012). Normally, perfect electric conductor (PEC) is taken as ground plane toward minimum losses. If the ground plane is included in the antenna design, the radiation waves from the antenna get reflected (scattered) by the ground plane. In fact, the ground plane acts as a perfect reflector for the incident EM waves. At low frequencies, if the ground plane is taken as a lossy material other than PEC, the incident waves will penetrate into the ground plane, inducing currents over the plane, resulting in ohmic losses. These losses in turn reduce the overall antenna efficiency.

The radiation pattern of the antenna also gets affected as all the incident waves from the antenna would not be reflected back by the ground plane (Yang et al. 2012). The antenna designer has to take care that the dipole should be at least $\lambda/4$ above the ground plane, in order to confine the radiation within the half space. In case of dipole array, the ground plane should be a good conductor and extend at least $\lambda/2$ beyond the feed points of the end dipoles of the array (Elliot 1981).

If the dipole array is over the curved platform, the curvature of the substrate would change the angle of incidence element to element. This would, in turn changes the current distribution in the dipole elements. Moreover, the current distribution in the dipoles is affected by mutual coupling, especially for closely spaced elements. The antenna performance depends on the radius of curvature of the array and the lattice spacing. The antennas mounted on singly curved surfaces, a type of conformal arrays are used in applications where a large angular coverage (azimuthal) is desired. These types of antennas include cylindrical and conical arrays.

In this book, the design and analysis of dipole array over planar and cylindrical substrates are carried out using the full-wave simulation-based package based on *finite element method* (FEM). It involves tetrahedron-shaped meshing with adaptive feature. When the model is designed and analyzed, the complete model is divided into tetrahedrons. The full-wave simulations are performed to calculate the antenna parameters like radiation characteristics, S-parameter, impedance, Y-parameters, gain, VSWR, Smith chart, field patterns, efficiency, etc.

This book is divided into five sections. Section 2 describes the design and performance analysis of a single half-wave center-fed dipole over planar and non-planar substrates and finite ground plane. The performance analysis of antenna is done in terms of input impedance, return loss, and radiation pattern.

The effect of both infinite and finite ground planes is analyzed. In Sect. 3, the design and performance analysis of linear dipole arrays of different configurations are presented. Results are shown for both planar and cylindrical substrates. Section 4 presents the design and analysis of planar dipole arrays on a cylindrical substrate. The study carried out is summarized in Sect. 5.

2   Single Dipole on Planar Ground Plane

In the present work, dipole (known to be the simplest) is taken as antenna element. The material for dipole is taken as PEC. Here, the EM design of dipole is carried out at 3 GHz, (i.e., $\lambda = 100$ mm). The dipole arms are assumed to be printed on the dielectric substrate. The performance of antenna depends on the constitutive parameters of substrate. The ground plane is taken as PEC, located 0.15λ below the substrate.

In general, the ground plane is placed $\leq 0.25\lambda$ from the dipole. This is due to the fact that the impedance looking into the ground plate from the printed dipole plane is an open-circuit condition. The ground plane serves as a return path for the current on the coplanar printed structures. Thus, for a compact dipole design, the extra ground plane area can be taken out for reducing the overall antenna dimensions. Moreover, the removal of extra ground plane will not affect the antenna performance because the current distribution is not concentrated in the ground plane. In addition, the direct field from the dipole and the reflected field from the ground plate constructively interfere, being in-phase (Daniil 1997).

2.1   EM Design

For EM design of a dipole, the Cartesian coordinate system is considered (Fig. 1). The dipole length is taken as 0.45λ. The antenna substrate is taken as cuboid ($1.45\lambda \times 1.025\lambda \times 0.011\lambda$), shown in Fig. 2.

The material of substrate is dielectric ($\varepsilon_r = 1.2$, $\tan\delta_e = 0.0009$, and $d = 0.011\lambda$). In order to make dipole over the substrate, the dipole arms and the feed gap are designed. The dimension of each dipole arm is ($0.222\lambda \times 0.025\lambda \times 0.0008\lambda$). The

Fig. 1 Cartesian coordinate
system used for EM design of
a dipole

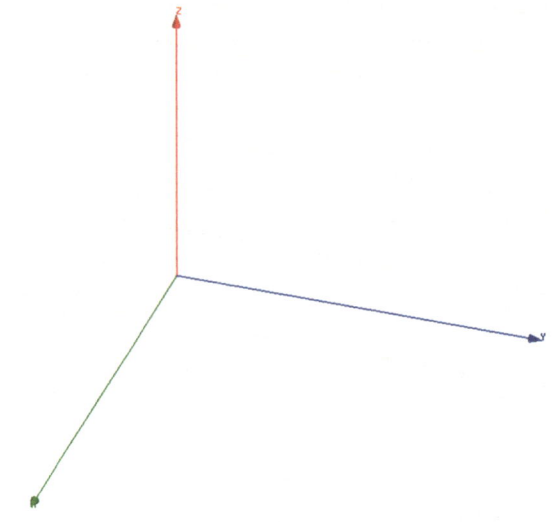

Fig. 2 Substrate for dipole
design

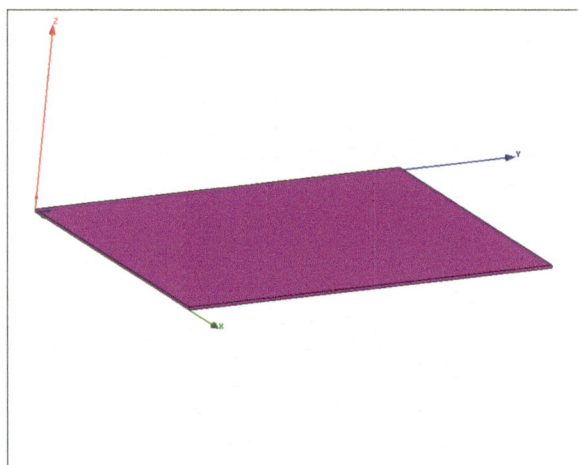

feed gap is rectangle strip ($0.006\lambda \times 0.001\lambda$). The dipole arms and the feed gap are taken as PEC (Fig. 3).

Once the dipole design is done, it needs to be excited for radiation. The lumped port excitation is given at the feed gap (from edge of one dipole arm to edge of the other dipole arm), as shown in Fig. 4. The multiple mode excitations are not considered. Other possible excitations are voltagesource, current source, wave port, plane wave excitations, etc. Next the ground plane is added at a distance of 0.15λ from the bottom face of the substrate (Fig. 5). The dimensions of the ground plane are the same as that of the substrate.

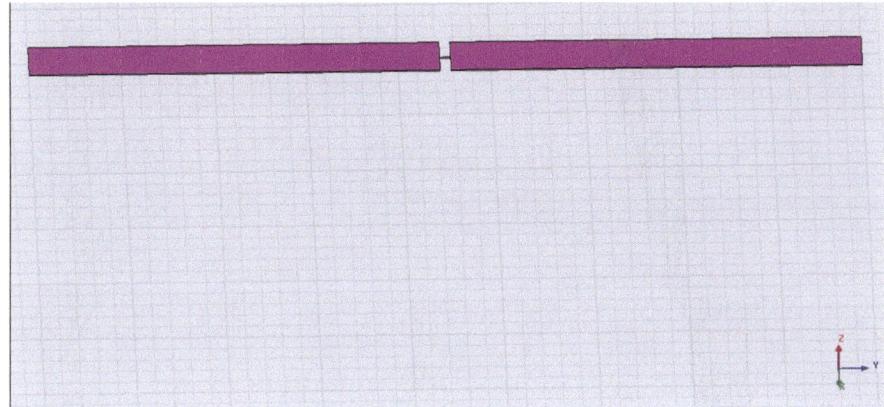

Fig. 3 Dipole arms and the feed gap on the substrate

Next step is to add radiation boundary in the design for simulating the radiation characteristics and other parameters like return loss, dipole impedance, etc. in far-field condition. The shape of radiation boundary can be cuboid, sphere, cylinder, or any other 3-D shape. The distance of radiation boundary should be at least $\lambda/4$ from the antenna element. Here, the distance between the dipole element and radiation boundary is taken as 0.25λ. Figure 6 shows the radiation boundary enclosing dipole with substrate. The radiation boundary is a cuboid ($1.45\lambda \times 1.025\lambda \times 0.025\lambda$).

2.2 Performance Analysis

The performance of dipole depends on the type of ground plane. Here, the performance analysis is discussed for both an infinite and a finite ground plane. Figure 7 shows the port field display at the dipole excitation. It may be observed that the dipole is excited properly. It shows the direction in which E-field propagates from one dipole arm to other through feed gap. The field distribution over the excited dipole antenna and the surface current over the dipole are shown in Figs. 8 and 9, respectively.

2.2.1 Infinite Ground Plane

In order to obtain the radiation pattern of the antenna in far-field, a perfect E-boundary condition is required to be imposed. This is done by including an infinite ground plane. Here the boundary is modeled as a finite portion of an infinite, perfectly conducting plane. In fact, a boundary condition in the form of an infinite

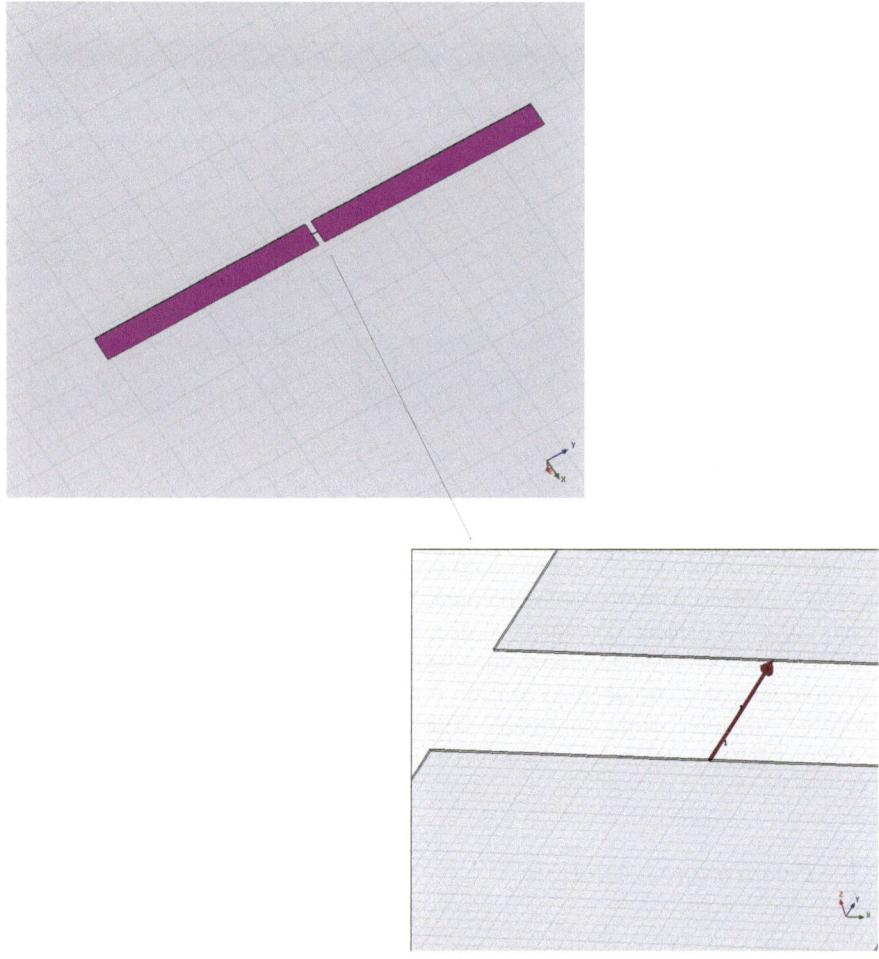

Fig. 4 Lumped port excitation at the feed gap of dipole

ground plane divides the design into two halves, one above the ground plane where the entire model resides, and other the half below it, where the radiated fields are set to zero.

An infinite ground plane must be defined on a planar surface. The design with curved surface cannot have infinite ground plane. The radiation pattern of a single dipole including infinite ground plane is shown in Fig. 10.

Figure 11 shows the corresponding real and imaginary parts of antenna impedance. It can be observed that the real part of impedance, $\text{Re}(Z) = 62.7638\Omega$ for $\text{Im}(Z) = 0$. Using this impedance value in the dipole design, the return loss obtained is shown in Fig. 12. The dipole resonant frequency is 2.752 GHz. The shift of resonant frequency from design frequency of 3 to 2.752 GHz is attributed to the

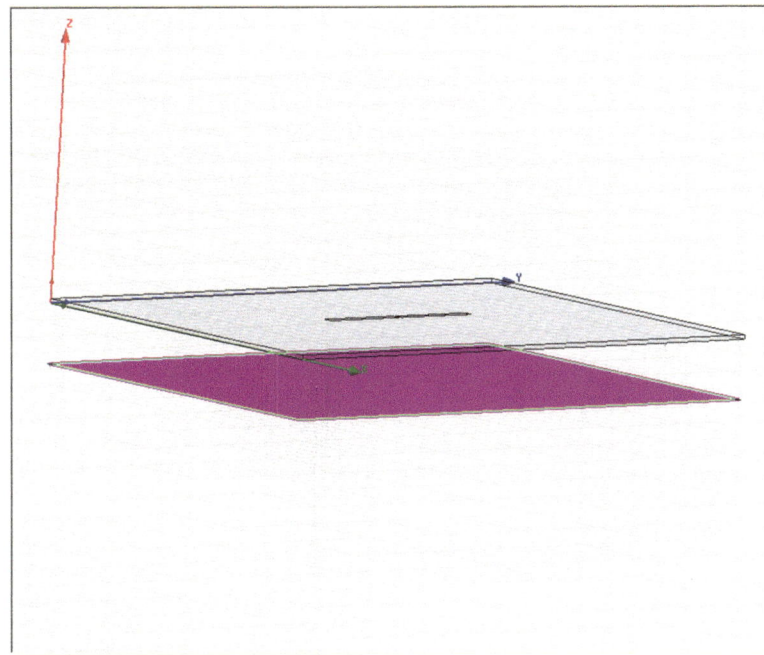

Fig. 5 Ground plane below the dipole and substrate

inclusion of dielectric substrate in the dipole design. This shift in resonant frequency depends on the constitutive parameters of the substrate. Higher the permittivity of the substrate material more will be the shift in resonant frequency. The corresponding 3-D radiation pattern of a single dipole antenna with an infinite ground plane is shown in Fig. 13.

2.2.2 Finite Ground Plane

In this section, the perfect E-boundary condition is imposed by making the ground plane finite. The ground plane is taken as rectangular sheet ($1.45\lambda \times 1.025\lambda$) at the height of -0.15λ, as shown in Fig. 5. The effect of finite ground plane is that the radiation pattern of the single dipole antenna gets changed. Figure 14 shows the radiation pattern of a single dipole antenna with finite ground plane of rectangular shape. The antenna impedance, Re(Z) = 62.7638Ω for Im(Z) = 0, as shown in Fig. 15. The return loss obtained is shown in Fig. 16. It can be observed that the resonant frequency obtained is 2.752 GHz. The corresponding 3-D radiation pattern is presented in Fig. 17. It can be observed that the pattern is cut below ground plane.

(a)

(b)

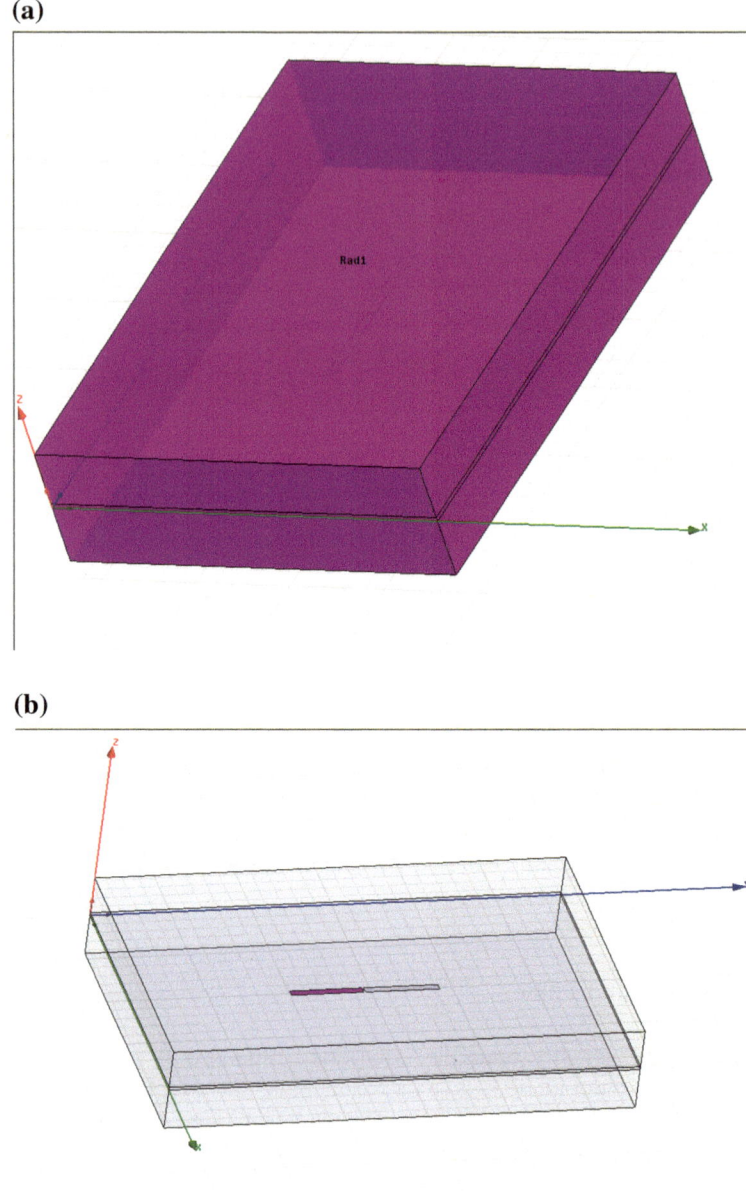

Fig. 6 Radiation boundary covering the dipole and the substrate

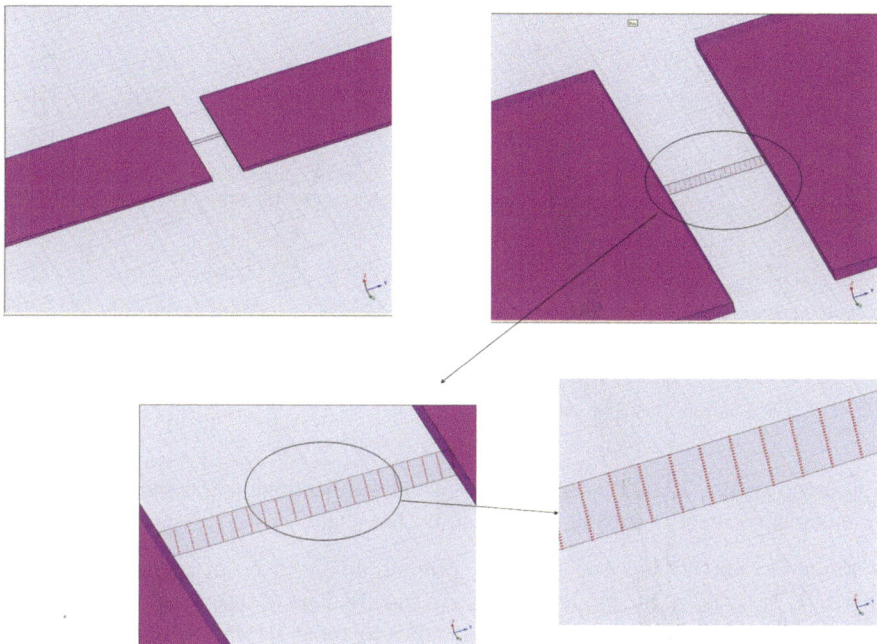

Fig. 7   Port field display at dipole excitation

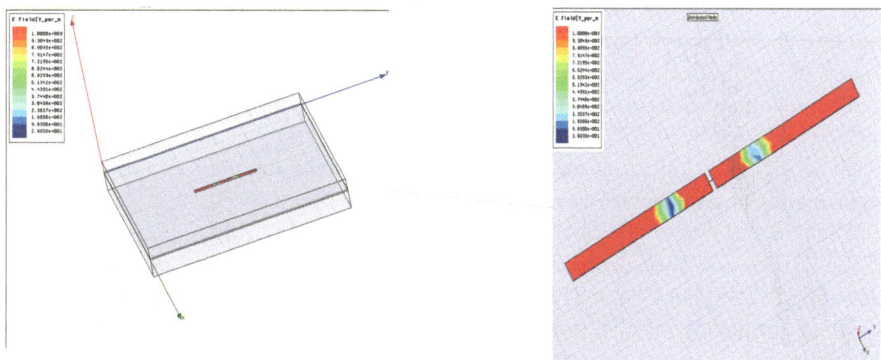

Fig. 8   E-field distribution over the dipole antenna

2.2.3   Cylindrical Substrate

The pattern synthesis of a conventional linear array is carried out based on the assumption of periodic identical antenna elements. This facilitates the analysis of array performance. However when an array is curved, the assumptions become invalid. The linear array methods like principle of pattern multiplication do not hold

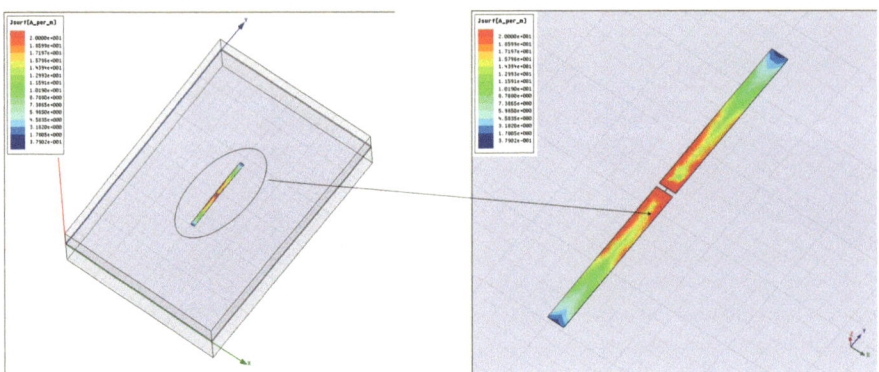

Fig. 9 Surface current over the excited dipole

for curved arrays. This is especially true for the applications where the placement of an antenna element is greatly influenced by the shape of the mounting platform.

Without Ground Plane For a single printed dipole over a cylindrical surface ($r = 0.9\lambda$, $h = 1.67\lambda$), the model is shown in Fig. 18. The design frequency of 3 GHz is considered. The thickness of the substrate is taken as 0.005λ. The dimensions of the dipole are taken as $l = 0.5\lambda$ including gap and $w = 0.003\lambda$. Rectangular sheet is taken as a feed gap (0.006λ and 0.0015λ). The dielectric constant and loss tangent of the substrate material are 1.5 and 0.0004, respectively. The dimensions of radiation boundary are $r = 1.15\lambda$ and $h = 1.92\lambda$. The excitation of the center-fed dipole is done through lumped port (Fig. 19). The port field display within the gap source of dipole is shown in Fig. 20.

The resultant E-field distribution over the dipole printed on the cylindrical surface is shown in Fig. 21. It can be observed that the field is uniformly distributed over the dipole, confirming the proper excitation of the dipole. The corresponding surface current distribution is shown in Fig. 22.

The real and imaginary parts of the antenna impedance over the frequency sweep (5–10 GHz) are shown in Fig. 23. The antenna impedance obtained is Re $(Z) = 66.20\Omega$ for Im(Z) = 0. The corresponding return loss is shown in Fig. 24. The resonant frequency of the printed dipole is obtained as 8.8 GHz. The shift in frequency is due to the dielectric substrate and the curvature of the substrate over which the dipole is printed.

The radiation pattern of a single dipole over cylindrical surface is shown in Fig. 25. Both the rectangular and polar plots for the radiation pattern at $\phi = 0°$ are shown. The corresponding 3-D radiation patterns are shown in Fig. 26.

With Ground Plane The design of a single printed dipole over a cylindrical surface shown in Fig. 18 is modified by adding a cylindrical ground plane (Fig. 27). The dimensions of hollow PEC cylindrical ground plane are ($r = 0.6\lambda$, $h = 2.5\lambda$,

(a)

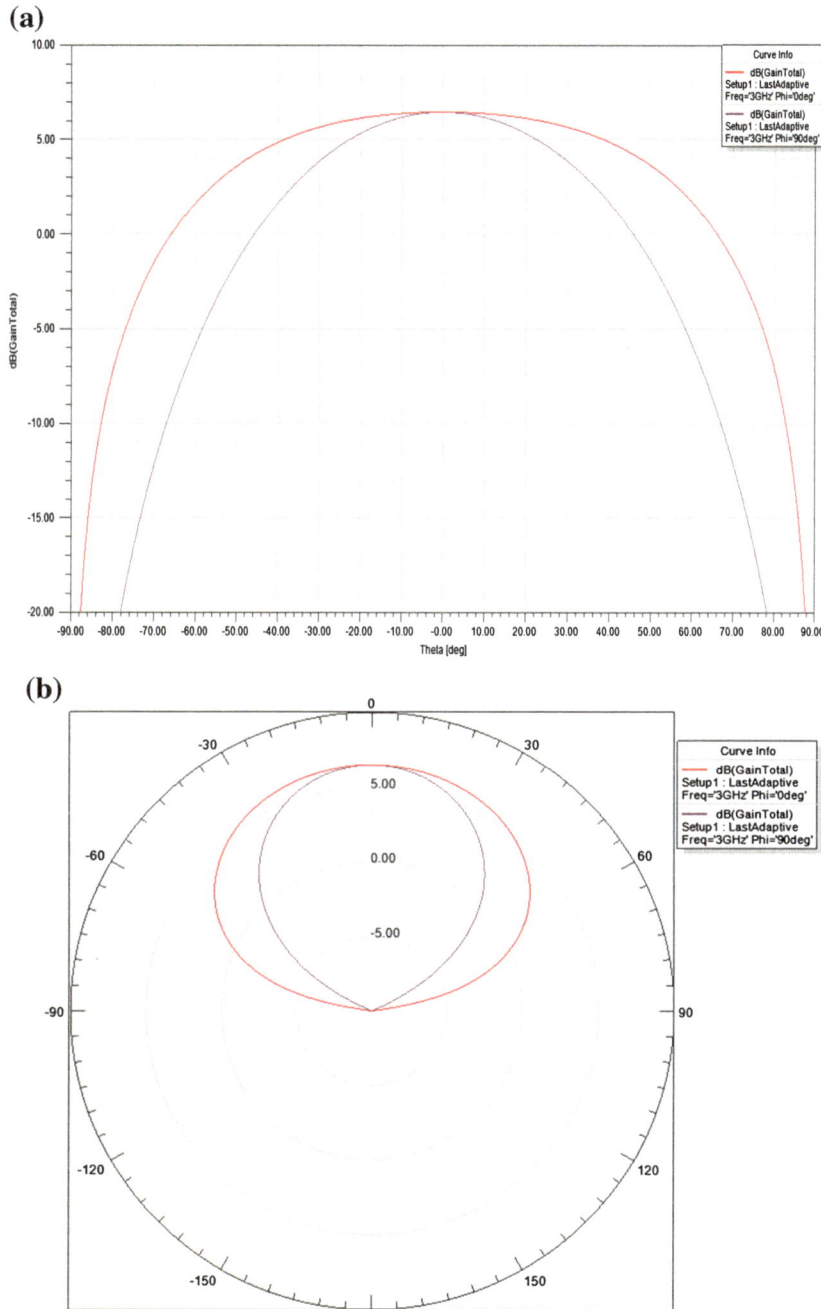

(b)

Fig. 10 Radiation pattern ($\phi = 0°$; $\phi = 90°$) of a single dipole antenna over an infinite rectangular ground plane. **a** Rectangular plot. **b** Polar plot

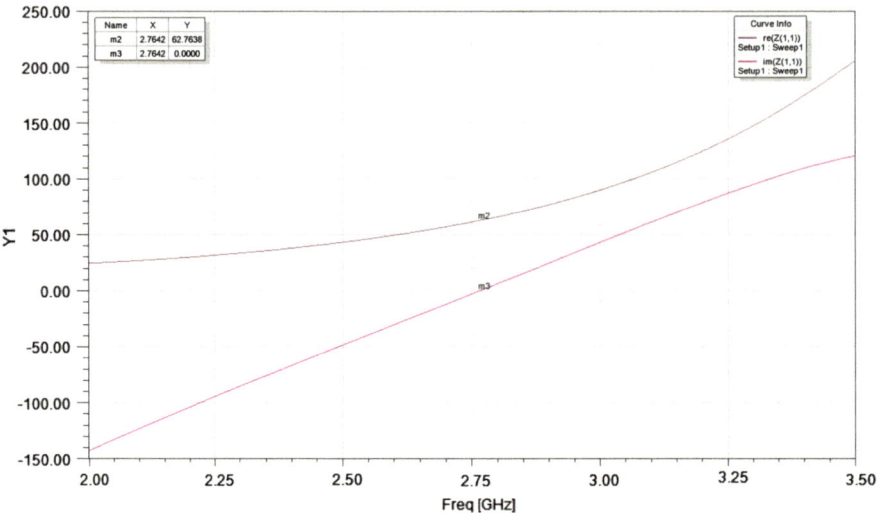

Fig. 11 Impedance of a single dipole antenna over an infinite ground plane

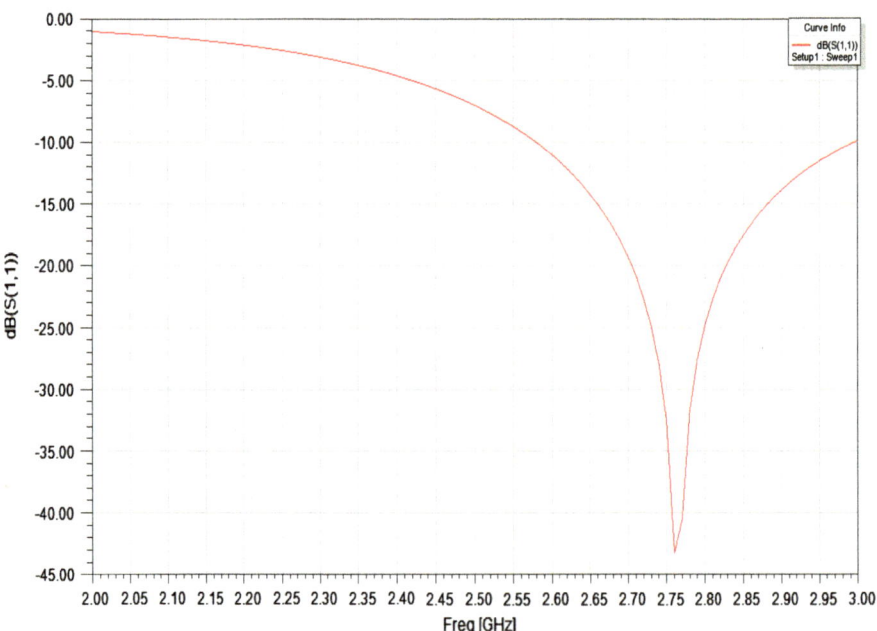

Fig. 12 Return loss of a single dipole antenna over an infinite ground plane

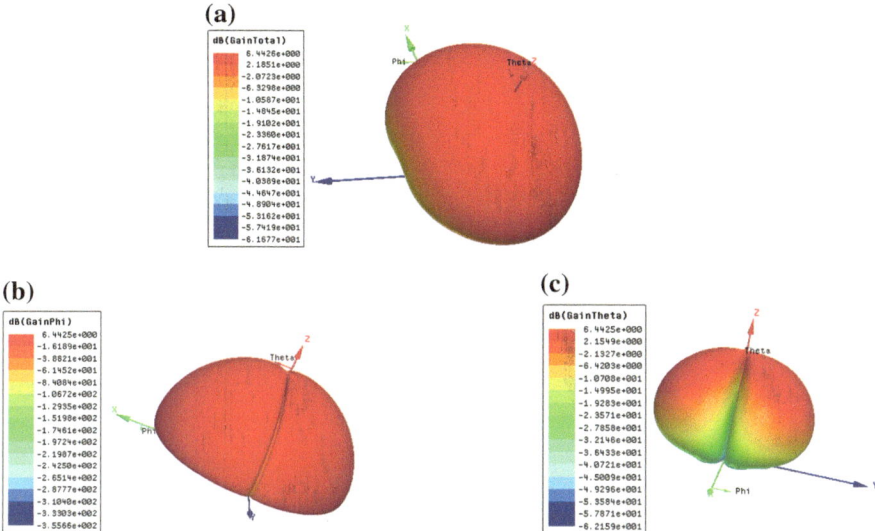

Fig. 13 3-D radiation pattern of a single dipole antenna over an infinite ground plane. **a** Total gain. **b** Gain (ϕ). **c** Gain (θ)

$t = 0.015\lambda$). The resultant impedance and return loss plots (Fig. 28) show that the antenna impedance is Re(Z) = 76.99Ω for Im(Z) = 0 and the dipole resonates at 2.6 GHz. The radiation pattern of the model is shown in Fig. 29. It can be observed that inclusion of ground plane in the design does not have any drastic change in the performance of a single dipole printed over a cylindrical substrate. However, the return loss got improved due to the ground plane. The design of a single printed dipole over a cylindrical surface shown in Fig. 27 is modified by truncating the ground plane, as shown in Fig. 30.

The corresponding impedance and return loss plot are shown in Fig. 31. The antenna impedance obtained is Re(Z) = 76.99Ω for Im(Z) = 0. The dipole resonates at 2.6 GHz, as shown in the return loss plot. The corresponding radiation pattern is shown in Fig. 32. It can be seen that the radiation pattern of the dipole slightly gets varied due to the truncation of the ground plane.

Figures 33 and 34 show the other variations of ground plane in a single dipole design. The overall antenna performance remains the same. However, the designer can choose any of these designs according to the application and the desired performance parameters like return loss or flatness of the main lobe in the radiation pattern.

(a)

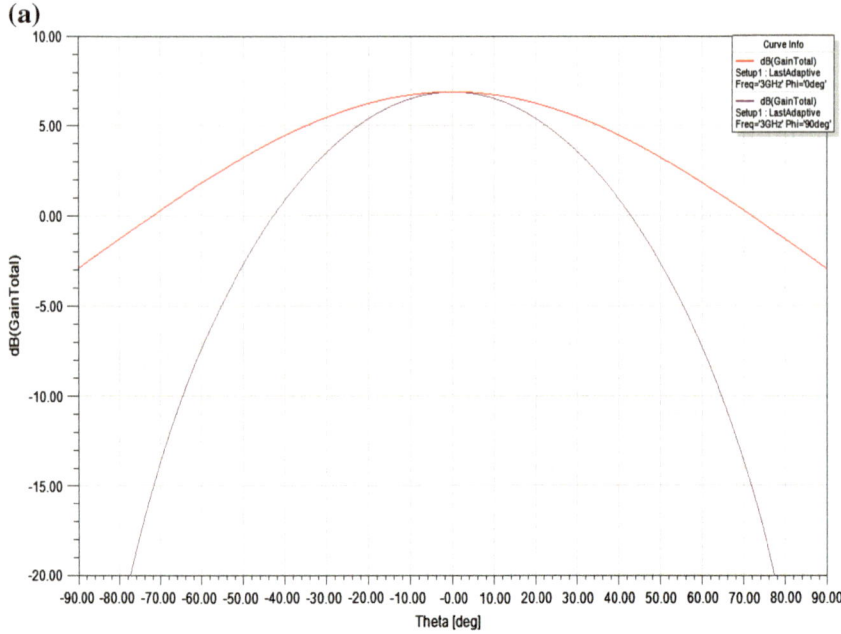

(b)

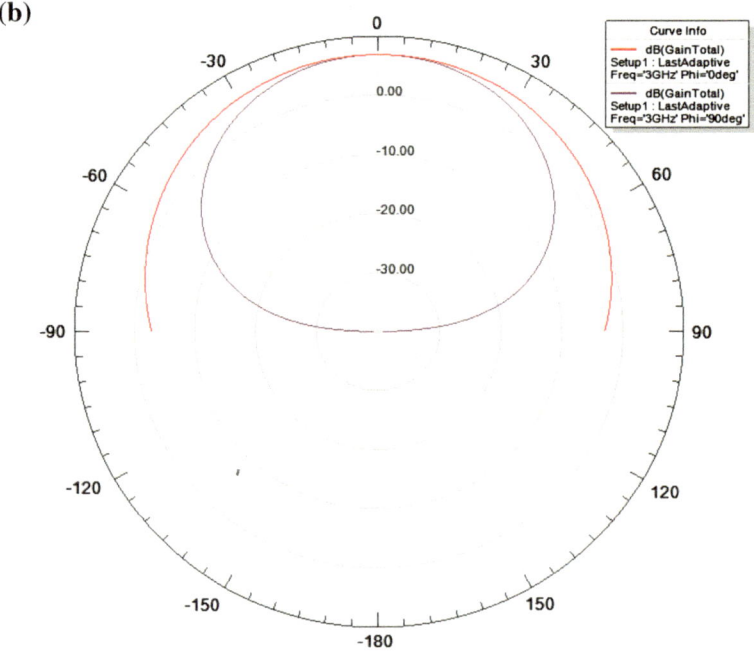

Fig. 14 Radiation pattern of a single dipole antenna over a finite rectangular ground plane. **a** Rectangular plot. **b** Polar plot

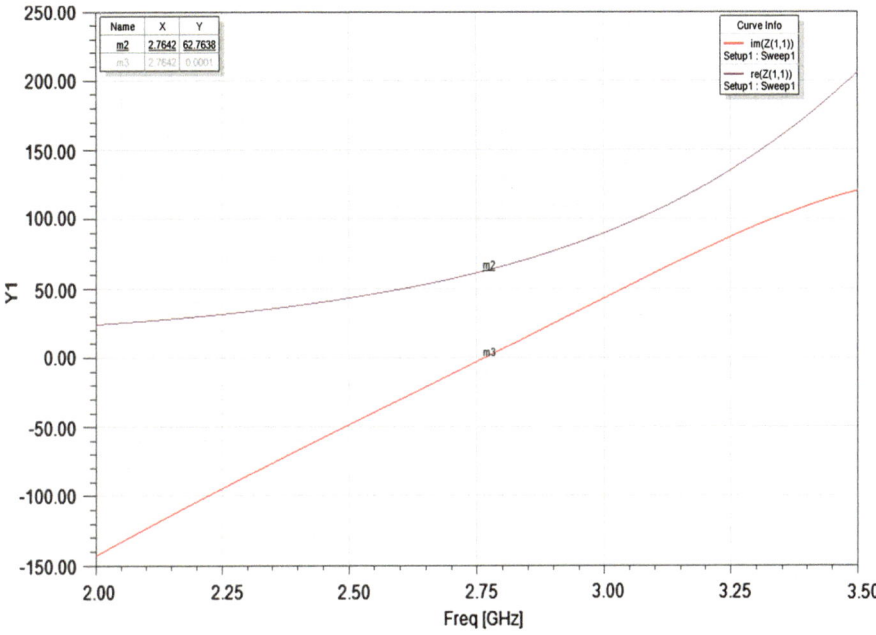

Fig. 15 Impedance (real and imaginary) plot of a single dipole antenna over a finite ground plane

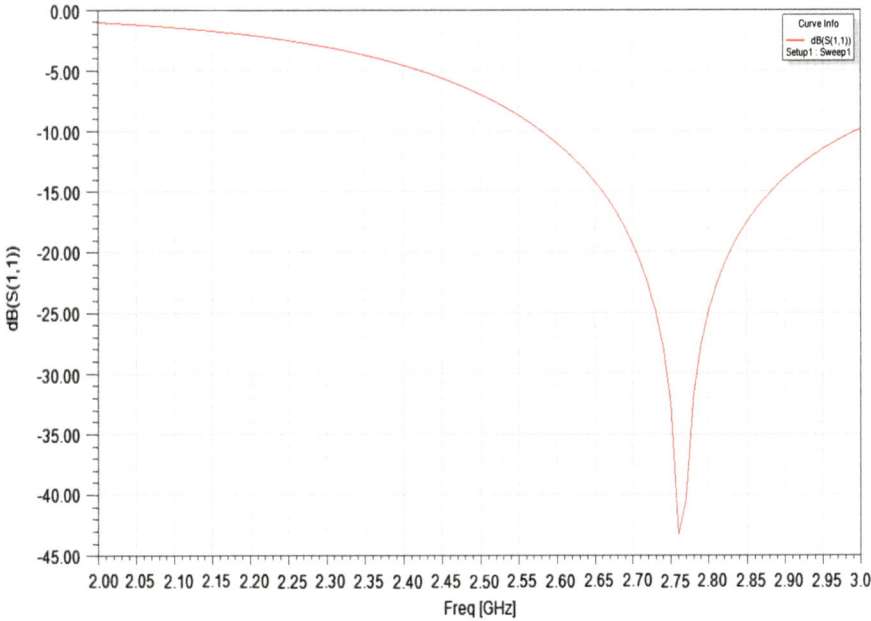

Fig. 16 Return loss of a single dipole antenna over a finite ground plane

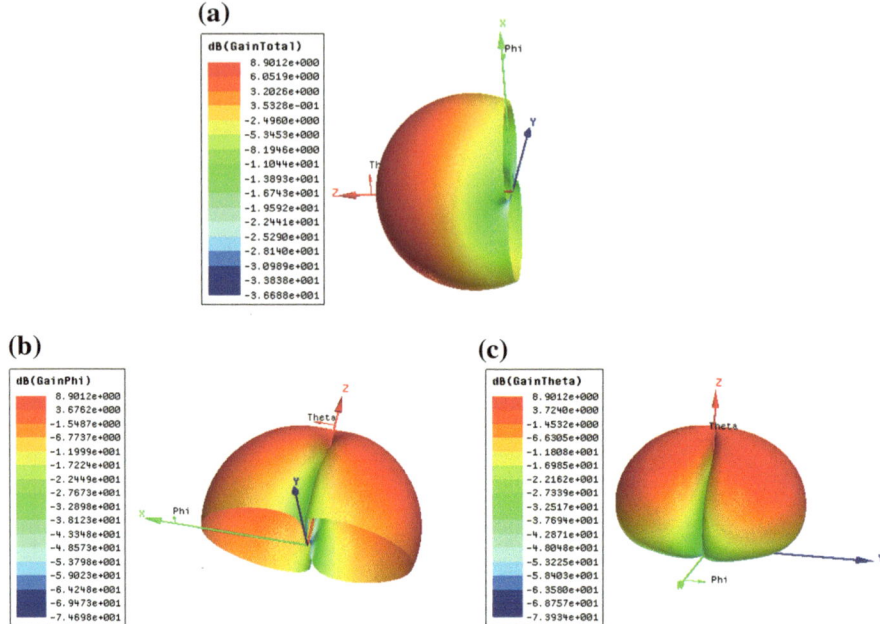

Fig. 17 3-D radiation pattern of a single dipole antenna over a finite ground plane. **a** Total gain. **b** Gain (ϕ). **c** Gain (θ)

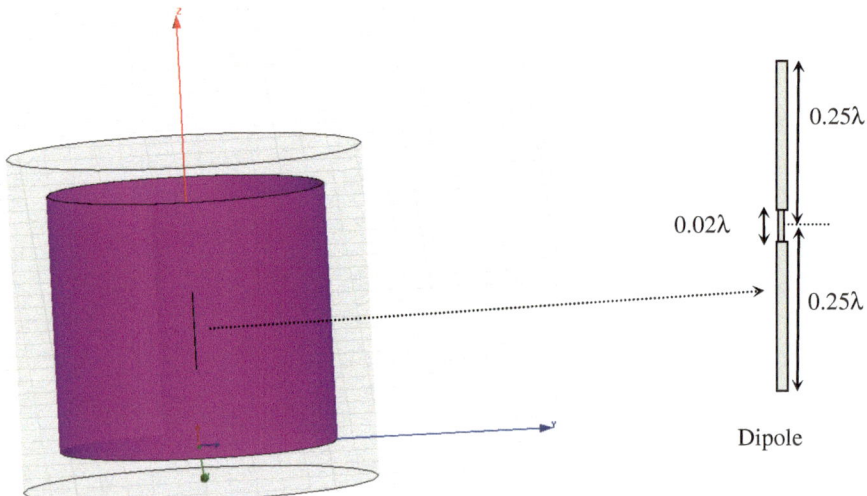

Fig. 18 A single dipole over a cylinder

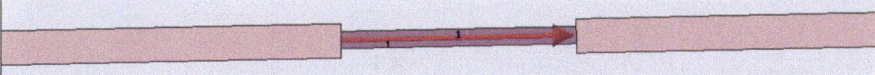

Fig. 19 Lumped port excitation at the feed gap of printed dipole

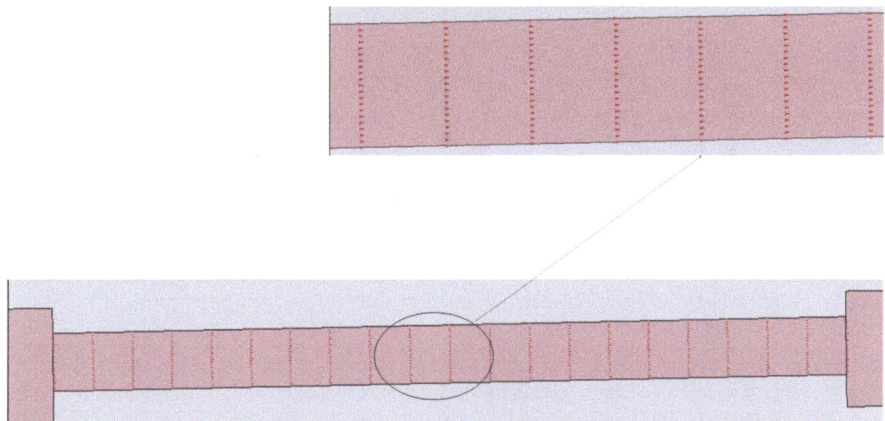

Fig. 20 Port field display at the feed gap of printed dipole

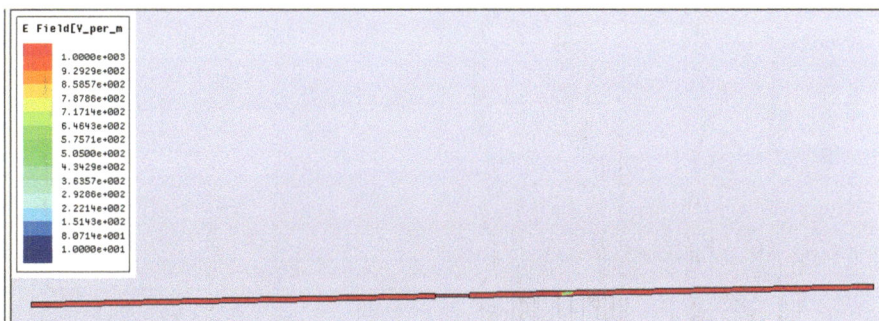

Fig. 21 E-field distribution over the printed dipole

3 Dipole Array Design

Next, the EM design center-fed half-wave dipole over planar and non-planar surfaces is done. Both linear and planar dipole arrays are designed and their performance is analyzed in terms of their return loss and radiation characteristics.

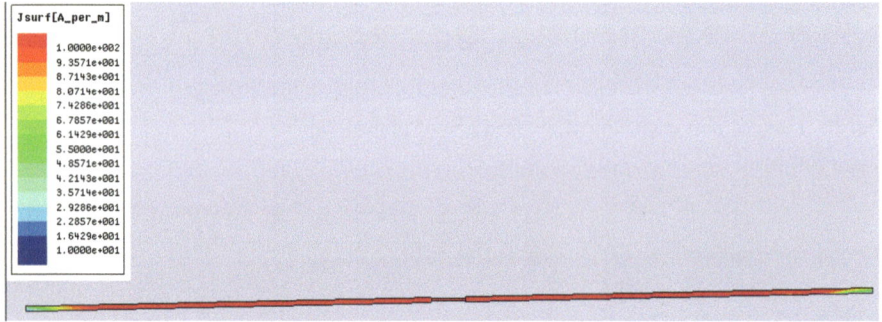

Fig. 22 Surface current over the printed dipole

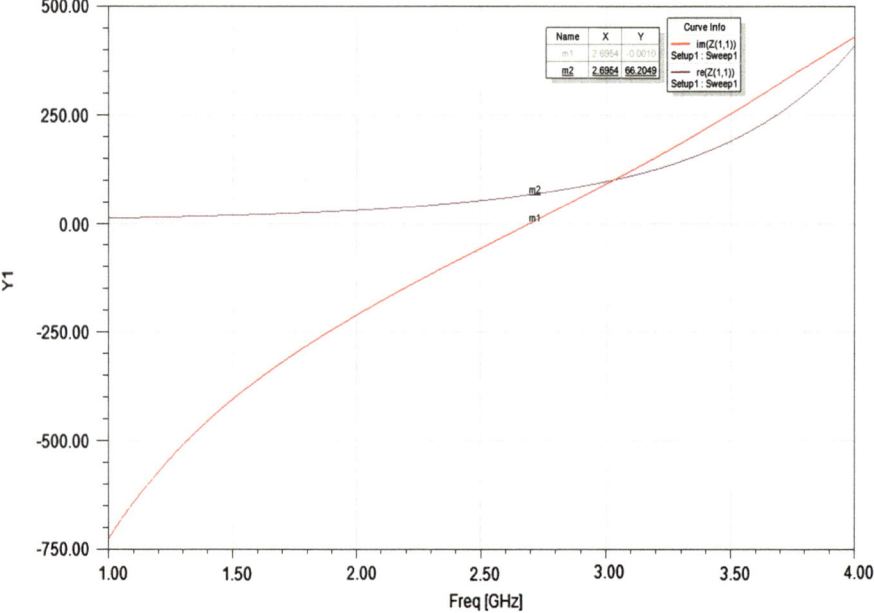

Fig. 23 Impedance (real and imaginary) of a single dipole antenna over a cylindrical surface

3.1 Two-Dipole Array Over Cylindrical Surface

The design parameters used are $r = 0.9\lambda$, $h = 1.67\lambda$, and the inter-element spacing over the cylindrical surface, $d = 0.55\lambda$. The design frequency is taken as 3 GHz. The resultant model is shown in Fig. 35. The truncated cylindrical ground plane is included in the design. All the other parameters are same as in Sect. 2.2.3. The excitation of both the center-fed dipoles is done through lumped port (Fig. 36).

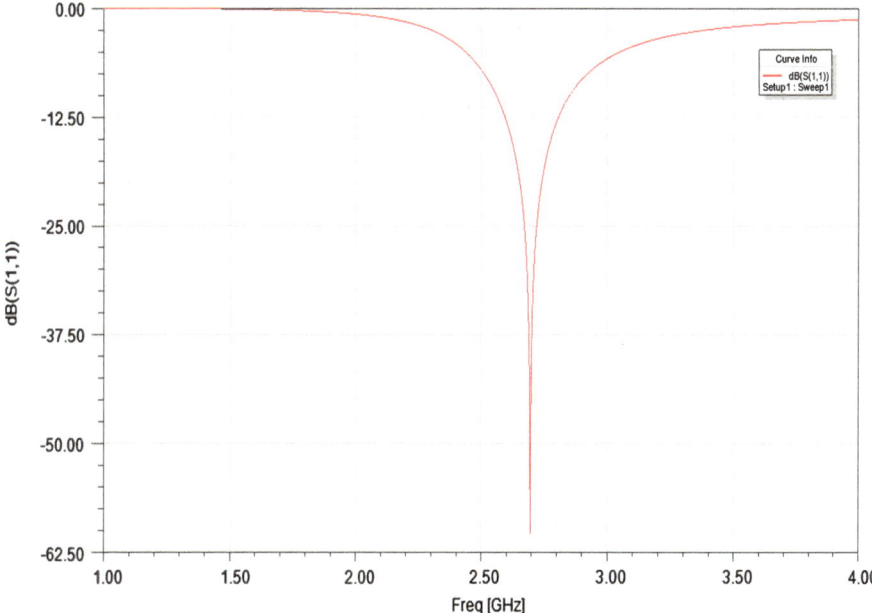

Fig. 24 Return loss of a single dipole antenna over a cylindrical surface

Since the substrate is curved, the inter-element spacing $d = 0.55\lambda$ corresponds to the arc length, s.

Arc length, $s = r\phi$.

Thus, $\varphi = s/r = 0.55\lambda/0.9\lambda = 0.61$

Using $\phi = 0.61$ radians, the coordinates of second dipole element on the substrate can be determined as follows:

$x = r\cos\varphi = 73.768$ mm; $y = r\sin\varphi = 51.558$ mm

One should note that first dipole element is assumed to be positioned at $\phi = 0$. Thus, for first dipole element, $x = r$; $y = 0$. Thus, for nth dipole, ϕ will be equal to $(n - 1)\phi$, where $\varphi = s/r$.

Table 1 presents the positions of the two dipoles printed on the cylindrical substrate.

The field port display at both the dipoles is shown in Fig. 37. It can be seen that both the dipoles are excited properly by lumped port. The corresponding E-field and the surface current distribution over the dipoles are presented in Figs. 38 and 39, respectively.

The impedance at the input terminals of the two dipoles over the cylindrical surface, over the frequency sweep (1–4 GHz) is shown in Fig. 40. The antenna impedance obtained is Re(Z) = 73.9083Ω; Im(Z) = 0. The corresponding return loss of the two dipoles is shown in Fig. 41. The dipoles are resonating at 2.51 GHz (approx.). The resultant radiation pattern of the two-dipole array design is shown in Fig. 42.

(a)

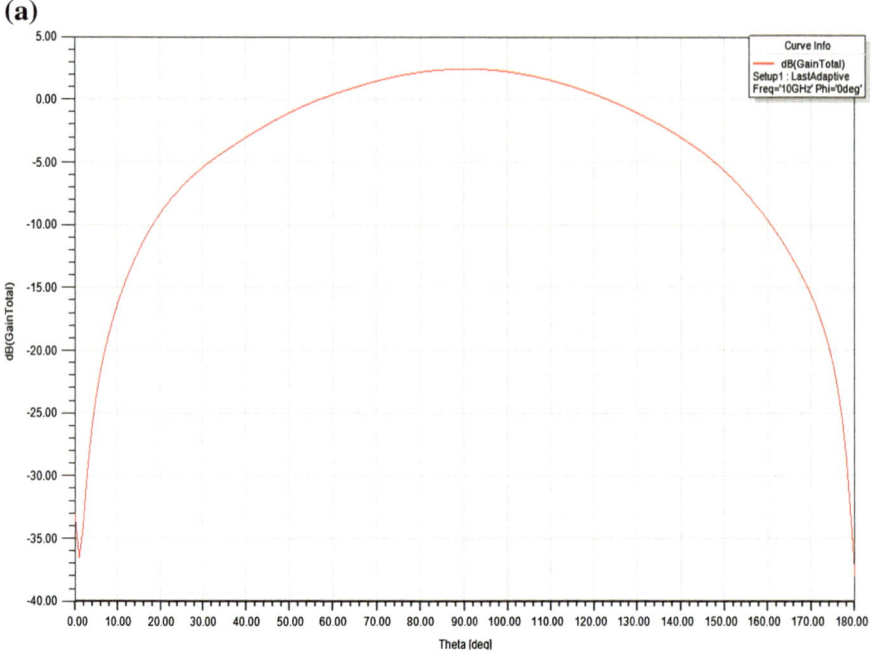

(b)

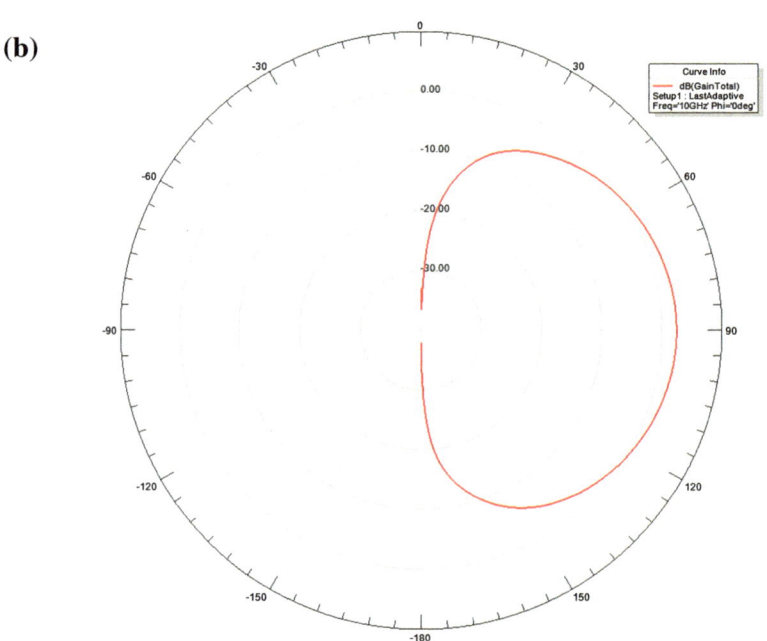

Fig. 25 Radiation pattern of a single dipole antenna over a cylindrical substrate. **a** Rectangular plot. **b** Polar plot

(a)   **(b)**

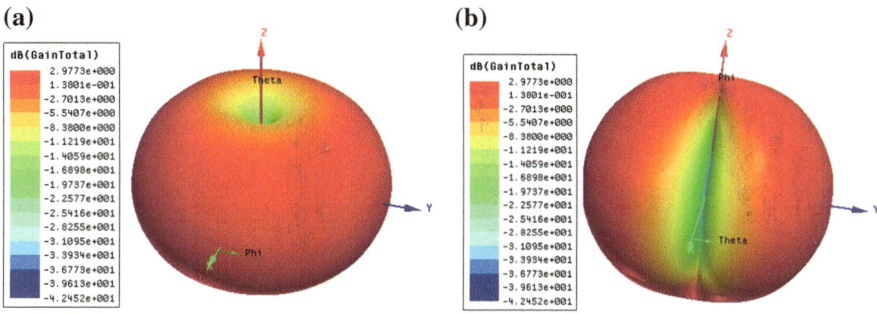

Fig. 26   3-D radiation pattern of a single dipole antenna over a cylindrical surface. **a** ϕ versus θ. **b** θ versus ϕ

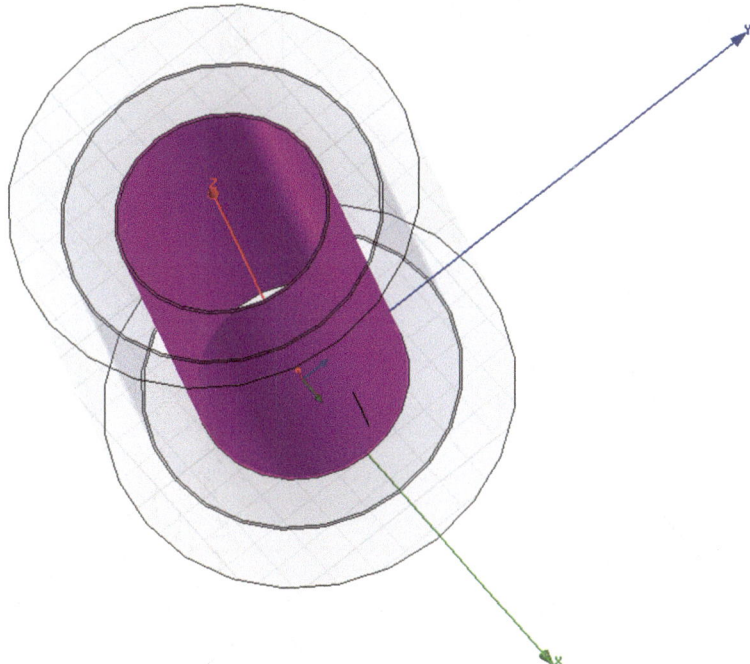

Fig. 27   A single dipole over a cylindrical surface with cylindrical ground plane

3.2   Three-Dipole Array

Next, the number of dipoles is increased to three. The dipole array is designed on planar substrate and cylindrical substrate. The constitutive parameters of substrate are taken as same as Sect. 2.

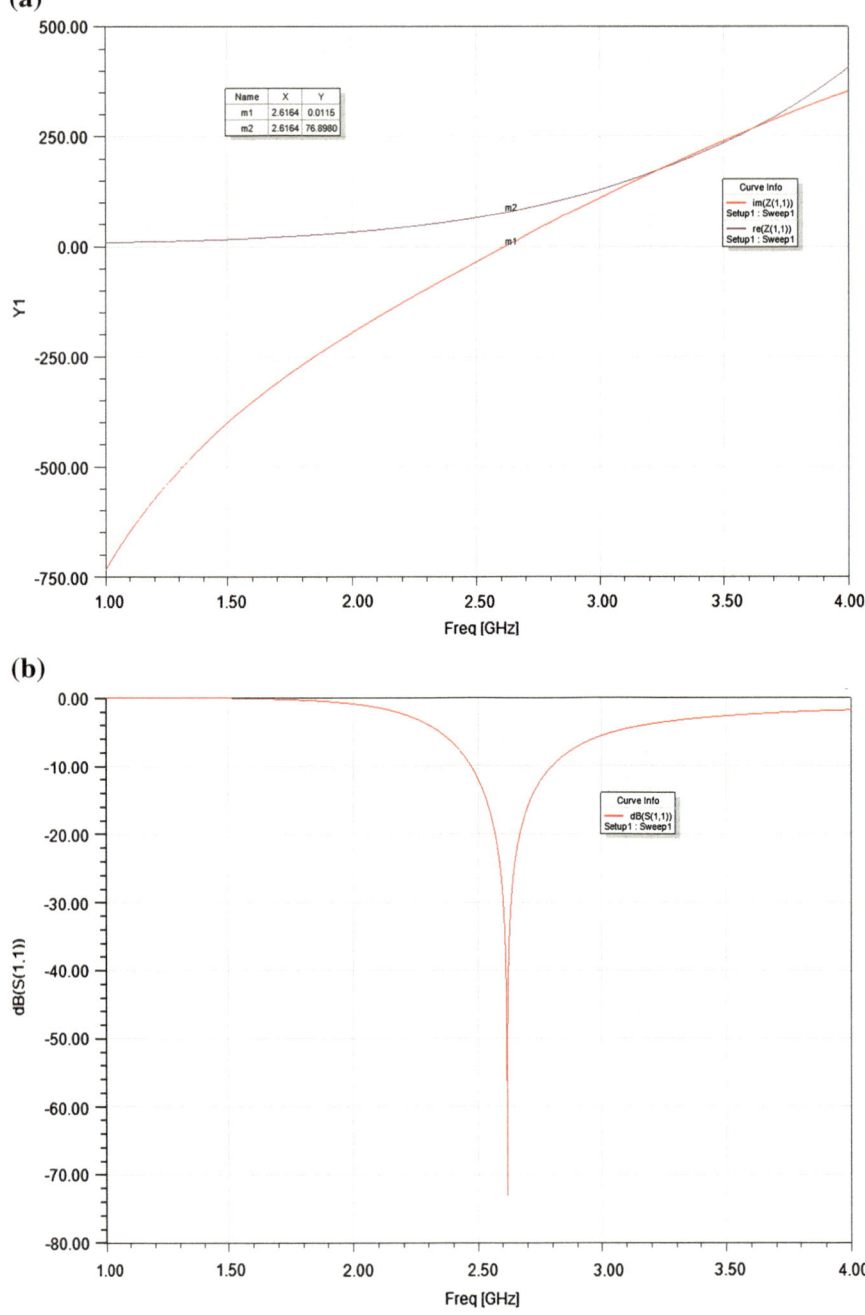

Fig. 28 Performance of a single dipole antenna over a cylindrical surface with cylindrical ground plane. **a** Impedance plot. **b** Return loss

Fig. 29 3-D radiation pattern
of a single dipole antenna
over a cylindrical surface with
cylindrical ground plane.
a Rectangular. **b** Polar. **c** 3-D

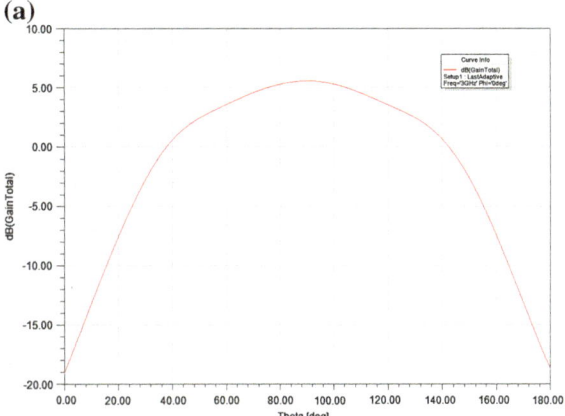

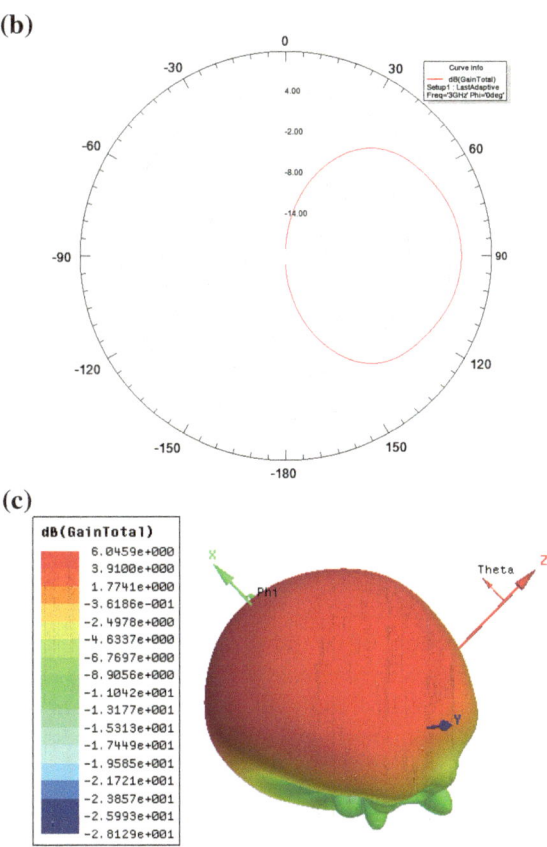

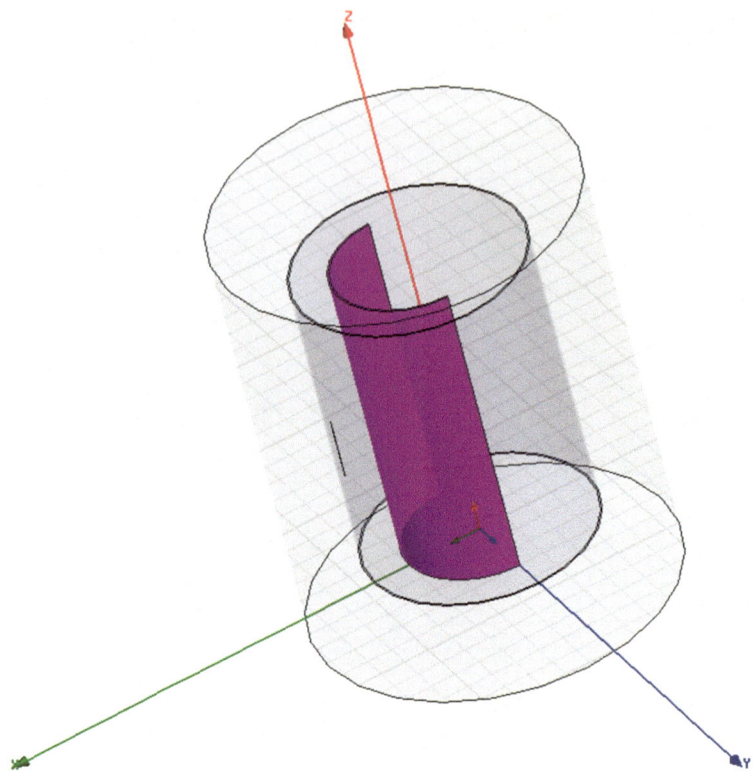

Fig. 30 A single dipole over a cylindrical surface with truncated ground plane

3.2.1 Planar Substrate

The substrate is taken as cuboid (2.575λ, 1.55λ, 0.05λ), as shown in Fig. 43. The spacing between the dipoles is 0.75λ. The vertical flap, connecting the dipoles is taken as a rectangular sheet (1.596λ × 0.00075λ). The radiation boundary is taken as cuboid of dimensions (2.575λ, 1.55λ, 0.5λ). The design is shown in Fig. 43. The ground plane is taken as rectangular sheet (2.575λ, 1.55λ) at the height of −0.15λ from the bottom of the substrate.

The dipoles are excited by the lumped port at the first dipole flap; the other two dipoles get excited due to the perpendicular flap joining the dipoles. The resultant E-field distribution over the three dipoles is shown in Fig. 44.

Infinite Ground Plane A perfect E-boundary condition is by including an infinite ground plane in the design, since the dipoles are over the planar surface. The radiation patterns obtained are shown in Fig. 45.

Finite Ground Plane On considering the finite ground plane, the radiation pattern of three-dipole array over planar substrate changes, as shown in Fig. 46.

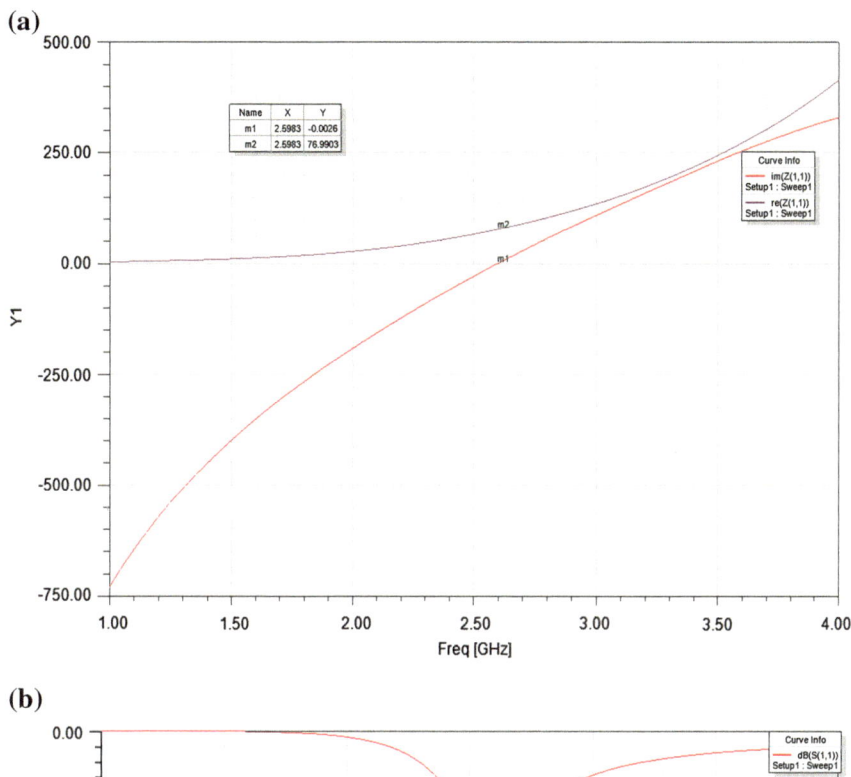

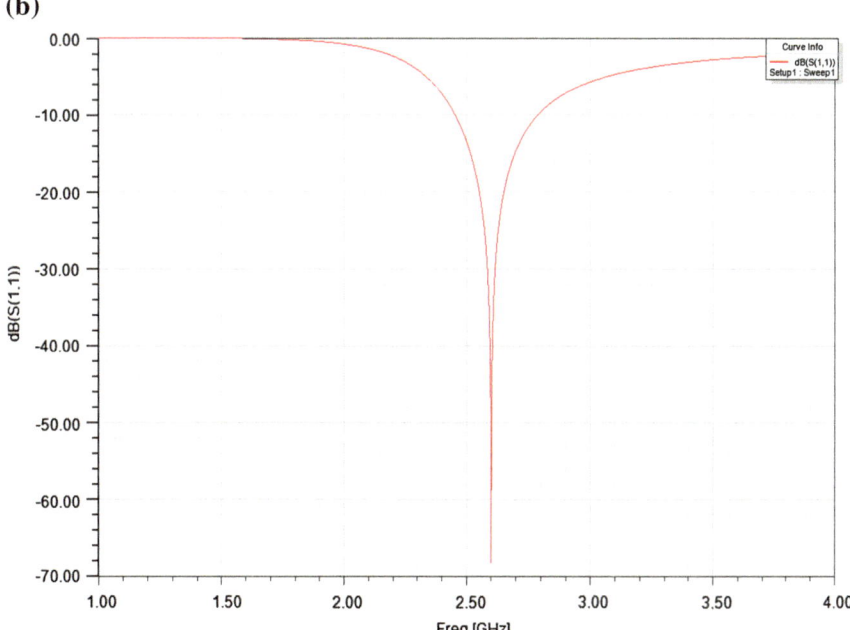

Fig. 31 Performance of a single dipole antenna over a cylindrical surface with truncated ground plane. **a** Impedance plot. **b** Return loss

(a)

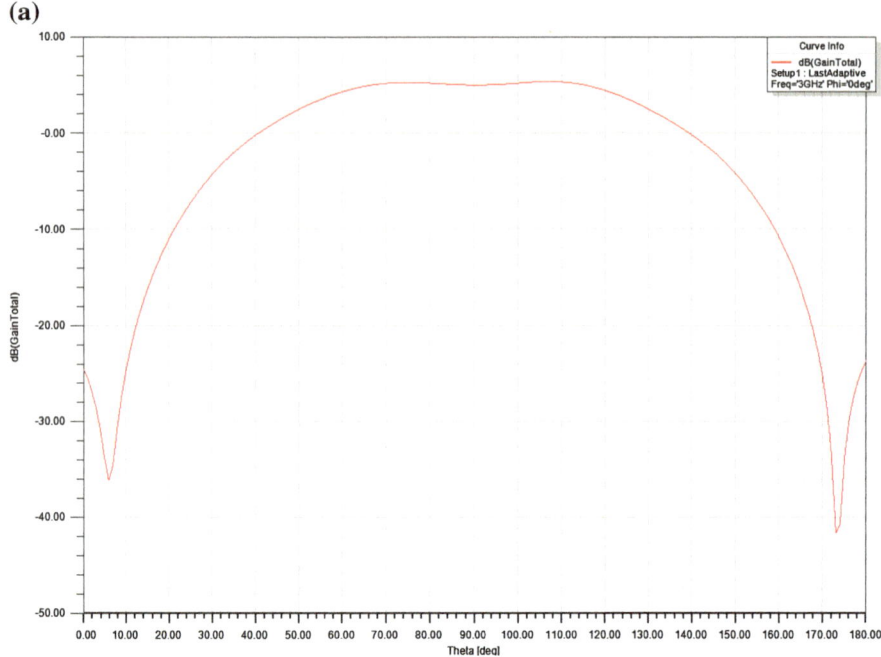

Fig. 32 3-D radiation pattern of a single dipole antenna over a cylindrical surface. **a** Rectangular. **b** Polar. **c** 3-D

3.2.2 Cylindrical Substrate

Here, the three-dipole array is designed on a cylindrical substrate with truncated ground plane. The design parameters used are same as in Sect. 3.1. The design is shown in Fig. 47. The excitation of the center-fed dipoles is done through lumped port. The dipoles are placed over the cylindrical substrate according to the coordinates, given in Table 2. The impedance at the input terminals of the dipole array over the cylindrical surface for the frequency sweep (1–4 GHz) is shown in Fig. 48. The antenna impedance obtained is $Re(Z) = 74.91\Omega$ for $Im(Z) = 0$.

The corresponding return loss of the dipole array is shown in Fig. 49. The dipole array resonates at 2.55 GHz. The radiation patterns (rectangular, polar, and 3-D) for three-dipole array are shown in Fig. 50.

(b)

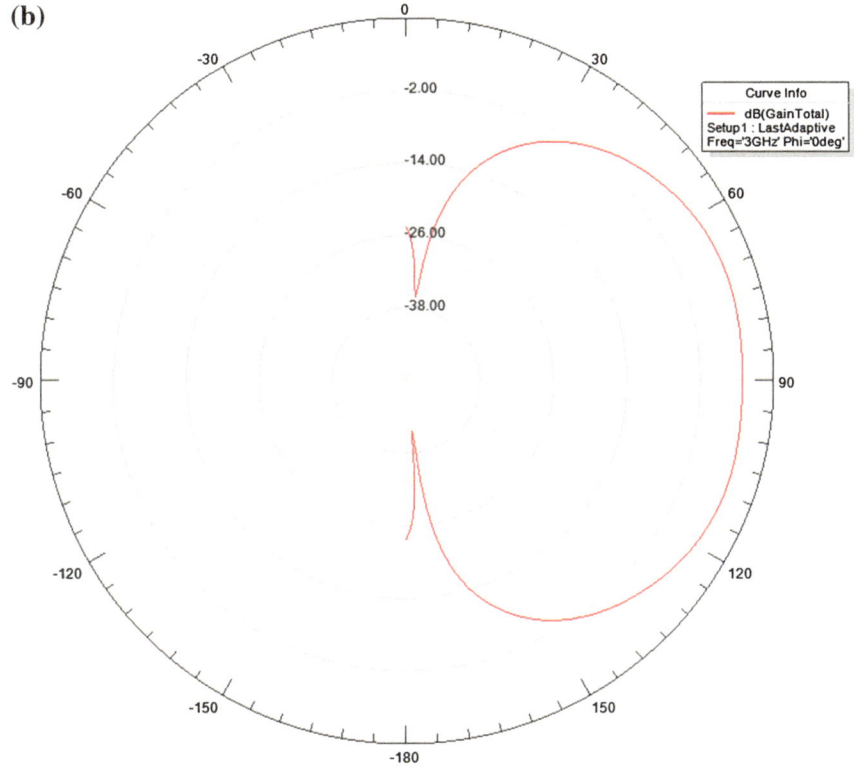

(c)

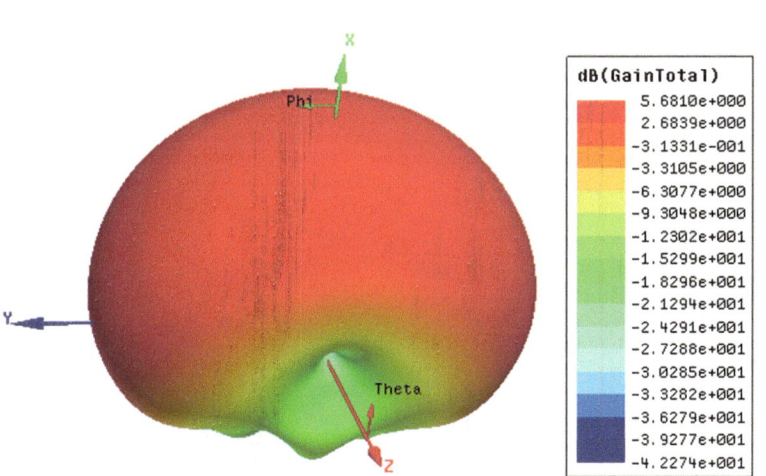

Fig. 32 (continued)

(a)

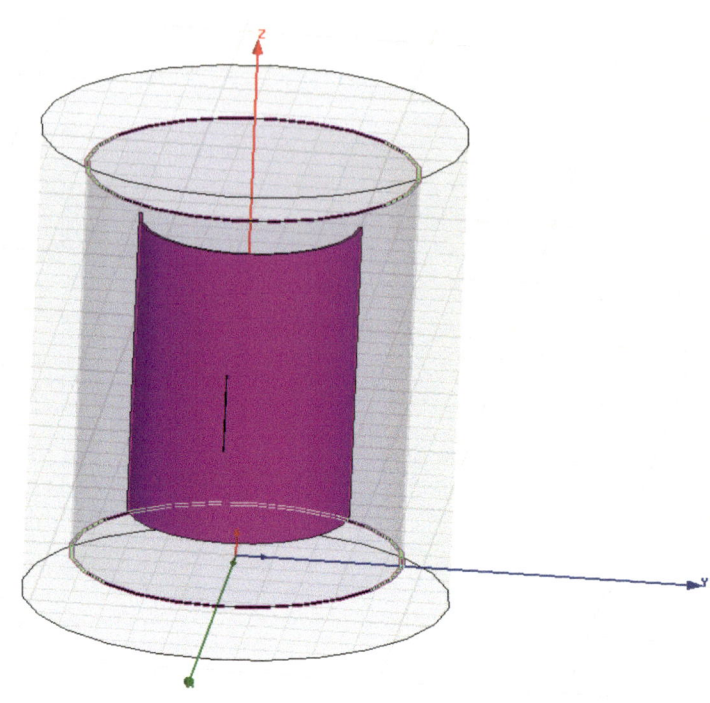

(b)

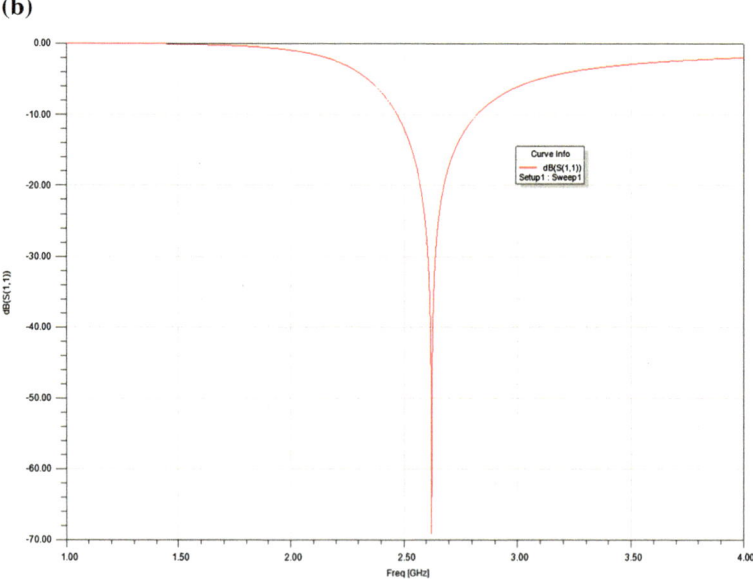

Fig. 33 Single dipole antenna over a cylindrical surface with truncated ground plane with reduced height. **a** Model. **b** Return loss. **c** and **d** Radiation pattern

(c)

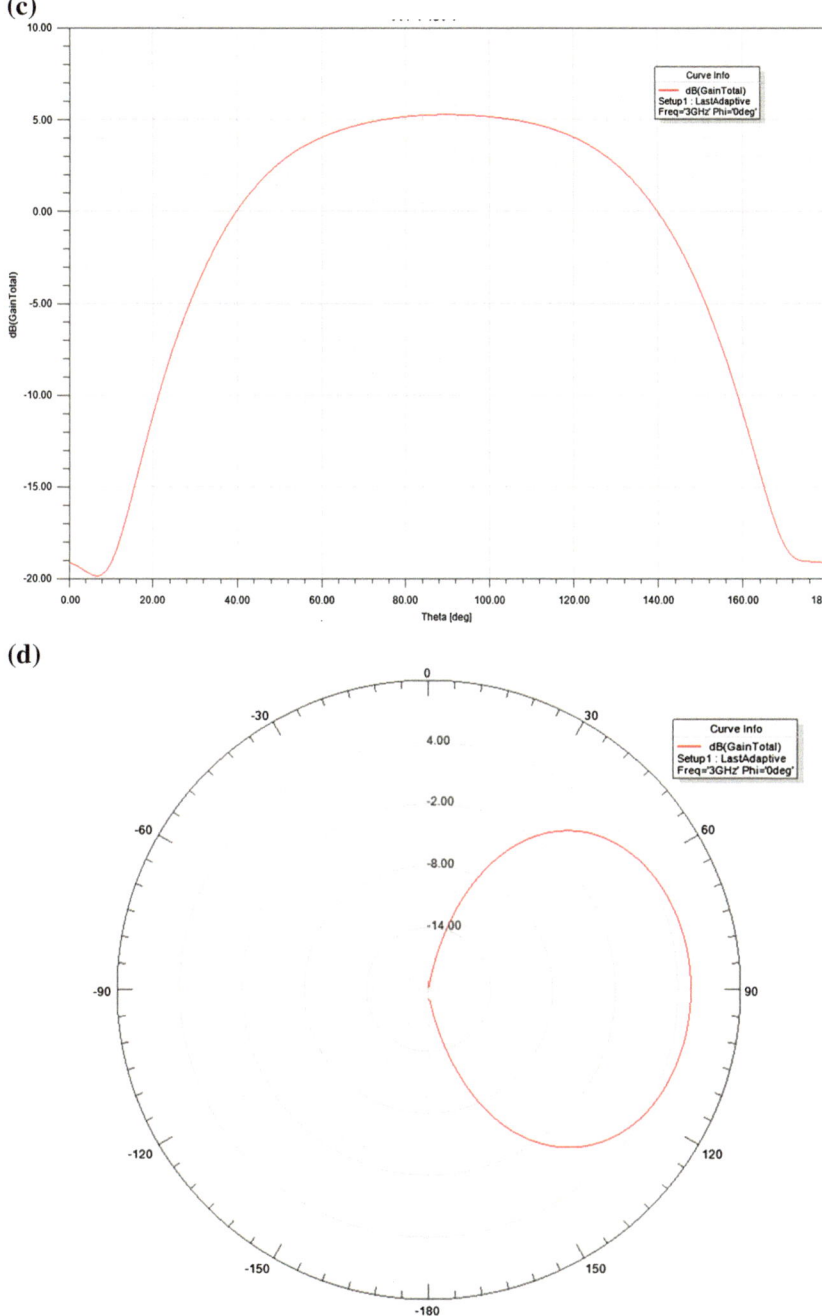

(d)

Fig. 33 (continued)

(a)

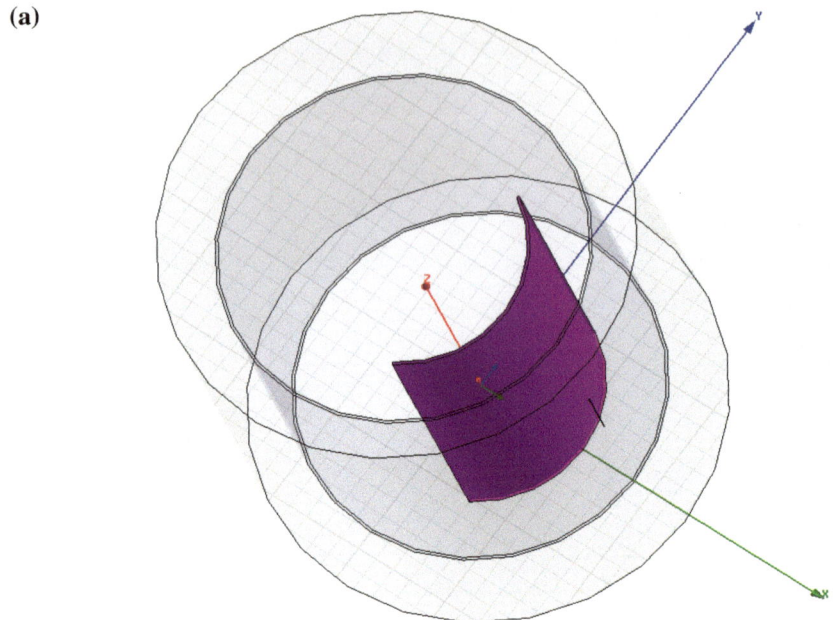

(b)

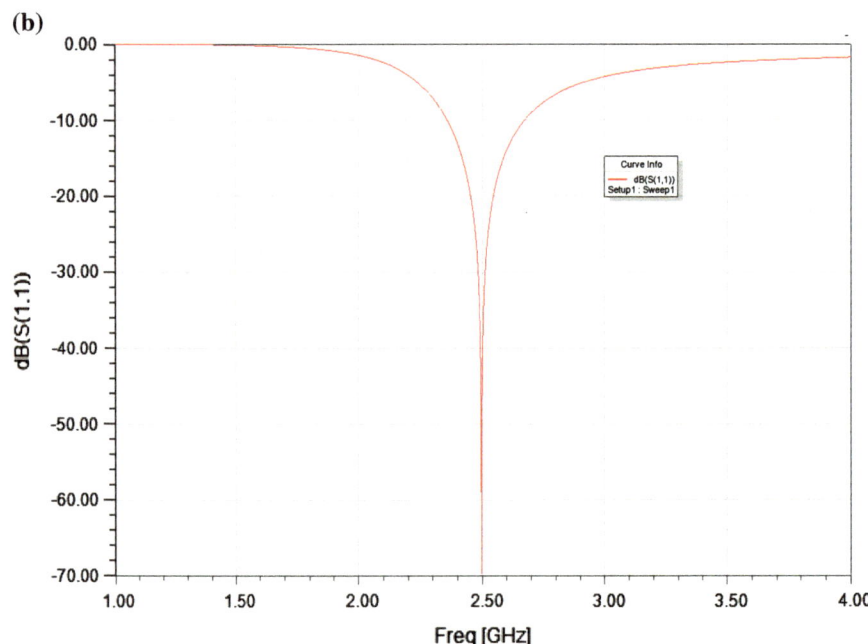

Fig. 34 Single dipole antenna over a cylindrical surface with sector (120°) ground plane. **a** Model. **b** Return loss. **c** Radiation pattern

(c)

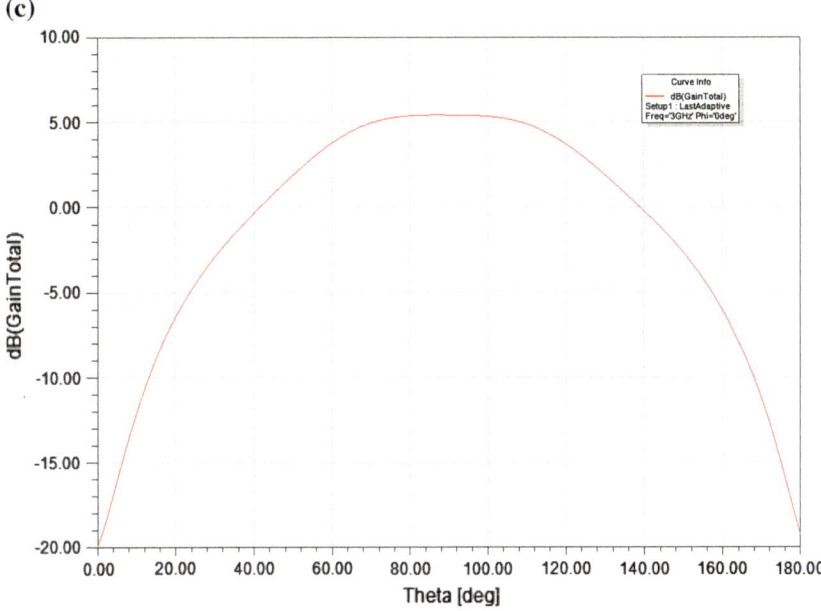

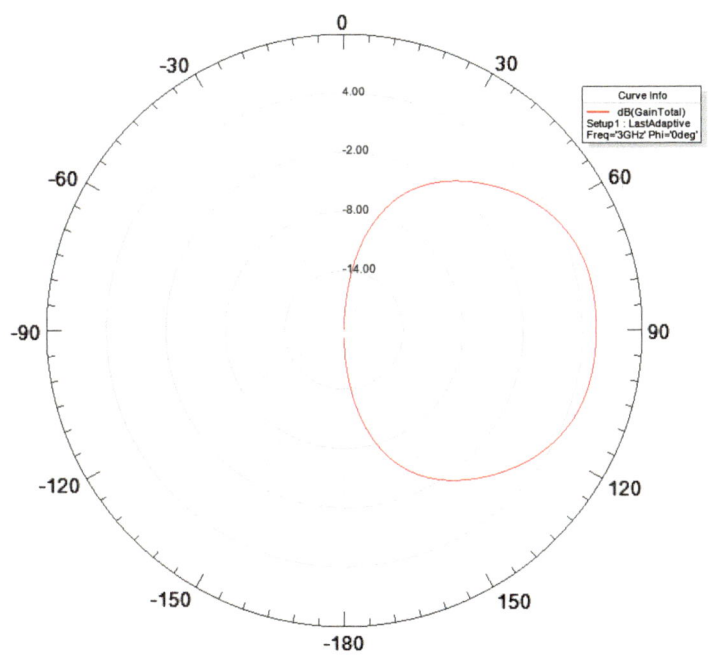

Fig. 34 (continued)

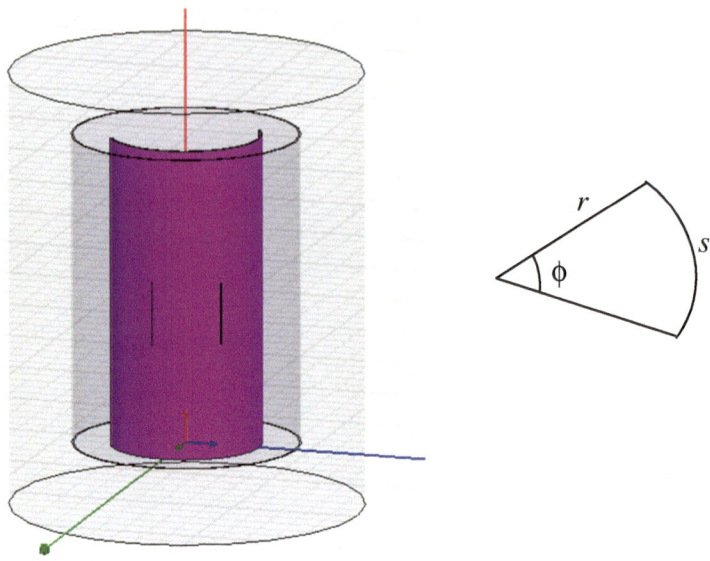

Fig. 35 Two-dipole array over a cylindrical surface with truncated ground plane

Fig. 36 Lumped port excitation of two-dipole array over a cylindrical surface with truncated ground plane. **a** Dipole 1. **b** Dipole 2

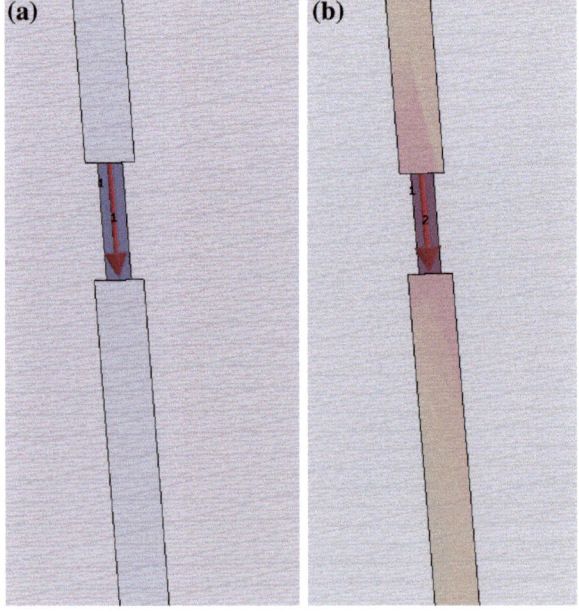

Table 1 Positions of two dipoles on the cylindrical substrate with truncated ground plane

Arc length, $s = 0.55\lambda$ and radius of cylindrical substrate, $r = 0.9\lambda$

Dipole	ϕ (radians)	Position of dipole on substrate	
		x-coordinate (mm)	y-coordinate (mm)
1	0	90	0
2	0.61	73.768	51.558

Fig. 37 Field port display of two-dipole array over a cylindrical surface with truncated ground plane. **a** Dipole 1. **b** Dipole 2

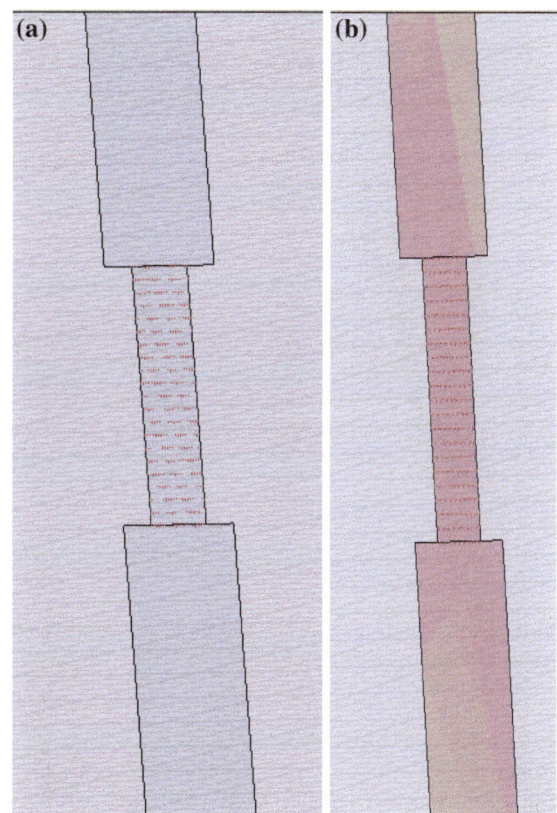

3.3 Five-Dipole Array

As above the number of dipoles is increased to five for planar and cylindrical substrates. The parameters are kept same as in Sect. 3.2.

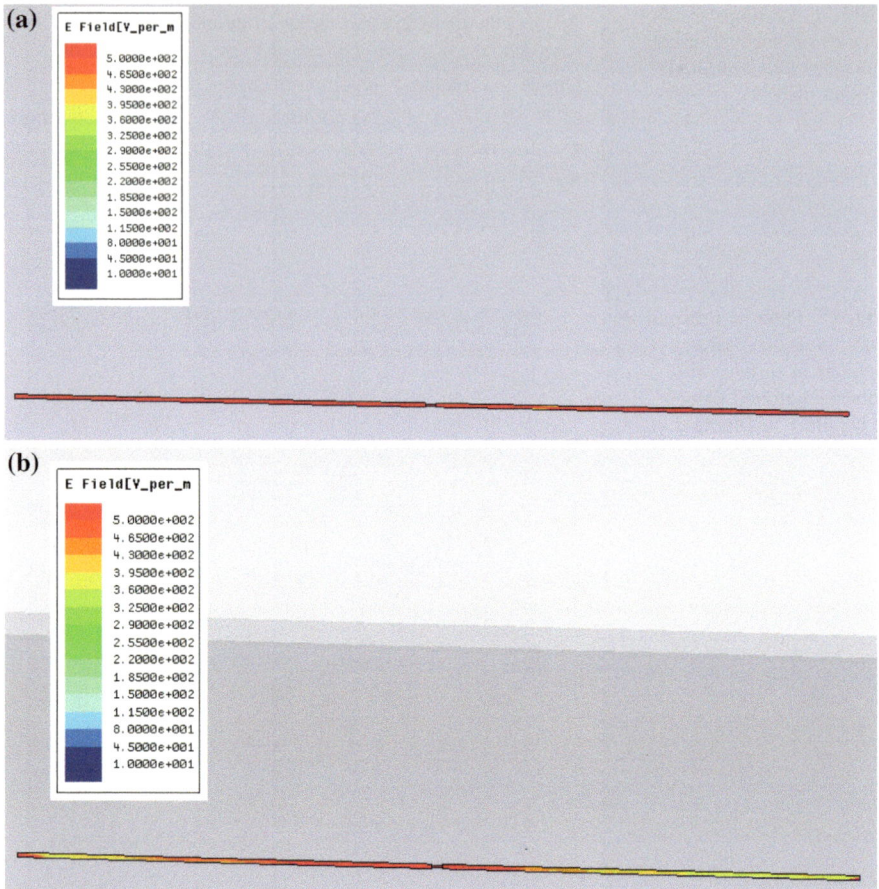

Fig. 38 E-field distribution of two-dipole array over a cylindrical surface with truncated ground plane. **a** Dipole 1. **b** Dipole 2

3.3.1 Planar Substrate

Since the number of dipoles is increased to five, the dimensions of vertical flap, substrate and ground plane, and radiation boundary get changed accordingly. The vertical flap, connecting the dipoles is taken as a rectangular sheet ($3.146\lambda \times 0.00075\lambda$). The dimensions of substrate are taken as (4.125λ, 1.55λ, 0.05λ), as shown in Fig. 51. The radiation boundary is of dimensions (4.125λ, 1.55λ, 1.75λ).

The ground plane is taken as rectangular sheet (4.125λ, 1.55λ) at the height of -0. 15λ from the bottom of the substrate. The dipoles, excited through lumped port have E-field distribution, shown in Fig. 52.

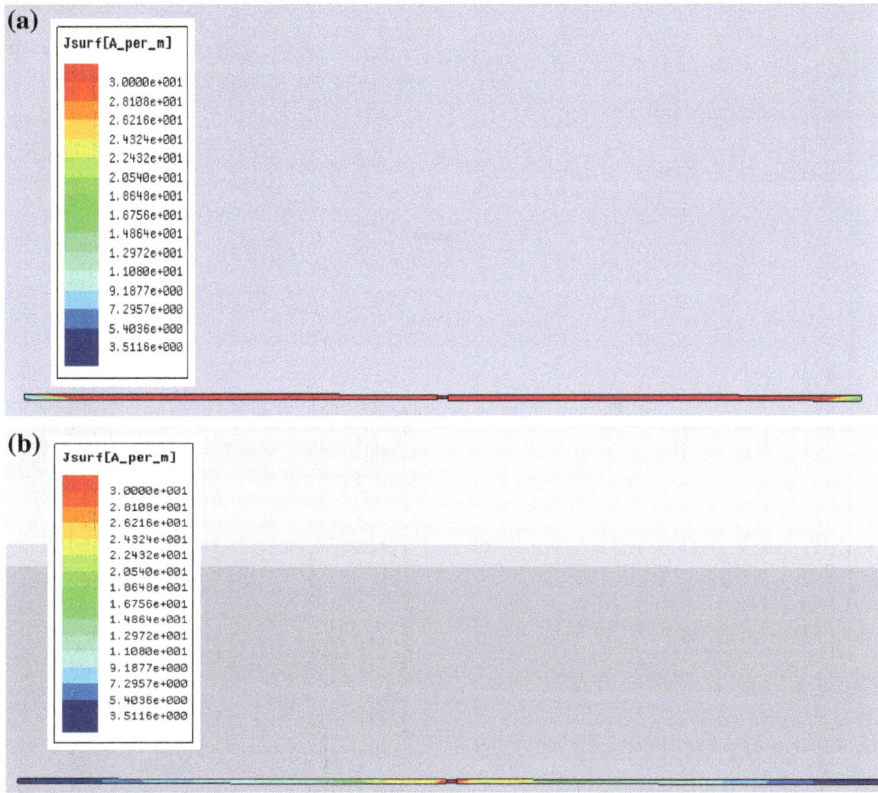

Fig. 39 Surface current distribution of two-dipole array over a cylindrical surface with truncated ground plane. **a** Dipole 1. **b** Dipole 2

Infinite Ground Plane The radiation pattern of five-dipole array over a planar substrate and infinite ground plane is shown in Fig. 53. It can be observed that the radiation pattern is different from that of previous designed dipole arrays. This is due to increase in the number of radiating elements, resulting in the narrow main lobe and occurrence of side lobes.

Finite Ground Plane Changing the ground plane into finite one, the radiation pattern changes a little, as shown in Fig. 54.

3.3.2   Cylindrical Substrate

Here, the five-dipole array is designed on a cylindrical substrate and truncated ground plane. The design parameters used are same as in Sect. 3.1.

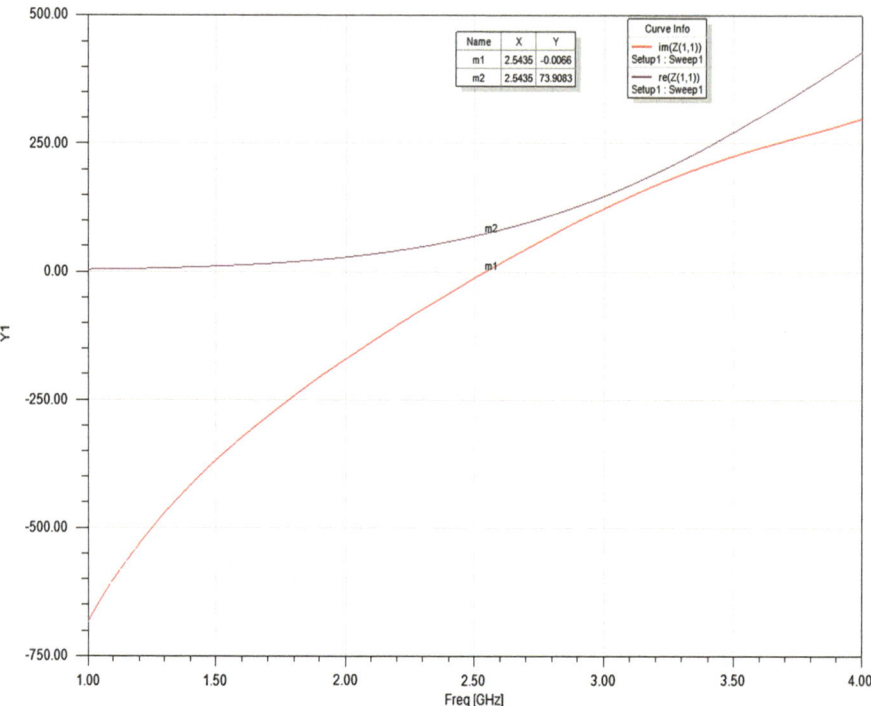

Fig. 40 Impedance (real and imaginary) plot of two-dipole array over a cylindrical surface with truncated ground plane

The design is shown in Fig. 55. The dipoles are placed on the cylindrical substrate according to the position coordinates, given in Table 3. The dipoles are excited using the lumped port. Figure 56 shows impedance plot, swept over the frequency range of 1–4 GHz.

The dipole antenna impedance is obtained as $Re(Z) = 73.53\Omega$ for $Im(Z) = 0$. The corresponding return loss of five-dipole array is shown in Fig. 57. It is apparent that the dipole array resonates at 2.5 GHz (approx). The design frequency is 3 GHz. The shift in resonant frequency from 3 to 2.5 GHz is due to the dielectric substrate. The radiation patterns of the 5-dipole array are shown in Fig. 58.

3.4   Ten-Dipole Array on Cylindrical Substrate

Next, the dipole array is designed with ten half-wavelength center-fed dipoles over a cylindrical substrate and hollow cylindrical ground plane ($r = 0.75\lambda$, $h = 2.5\lambda$, $t = 0.015\lambda$). The design parameters like dipole dimensions, flap dimensions, inter-element spacing, substrate material, thickness, etc., are same as in Sect. 3.1.

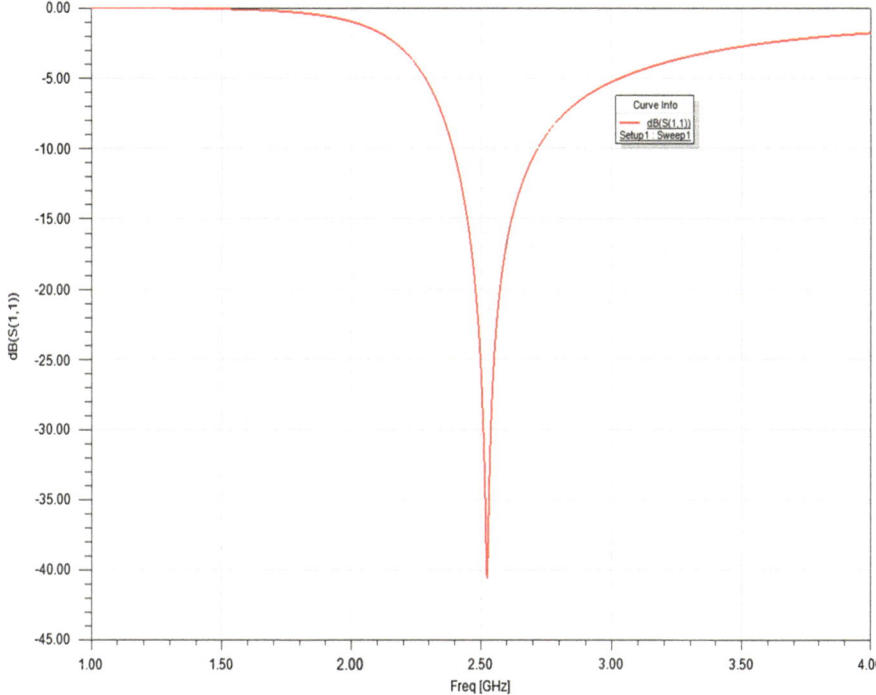

Fig. 41 Return loss of two-dipole array over a cylindrical surface with truncated ground plane

The design model is shown in Fig. 59. Using the coordinates given in Table 4, the dipoles are placed on the cylindrical substrate. The design frequency used is 3 GHz. Excitation of dipoles is done through lumped ports.

The variation of real and imaginary parts of antenna impedance of a ten-dipole array over a cylindrical substrate, for frequency sweep of 1–4 GHz is shown in Fig. 60. It can be observed that the antenna impedance at the input terminals is Re $(Z) = 28.97\Omega$ for Im(Z) = 0.

The resonance frequency of the array is obtained as 2.42 GHz (approx) from the return loss, shown in Fig. 61. The radiation pattern of a ten-dipole array is shown in Fig. 62. It is apparent that due to increase in the number of dipole elements in the array over the cylindrical substrate, the radiation characteristics have changed. It is apparent that the pattern is much different from other arrays designed so far.

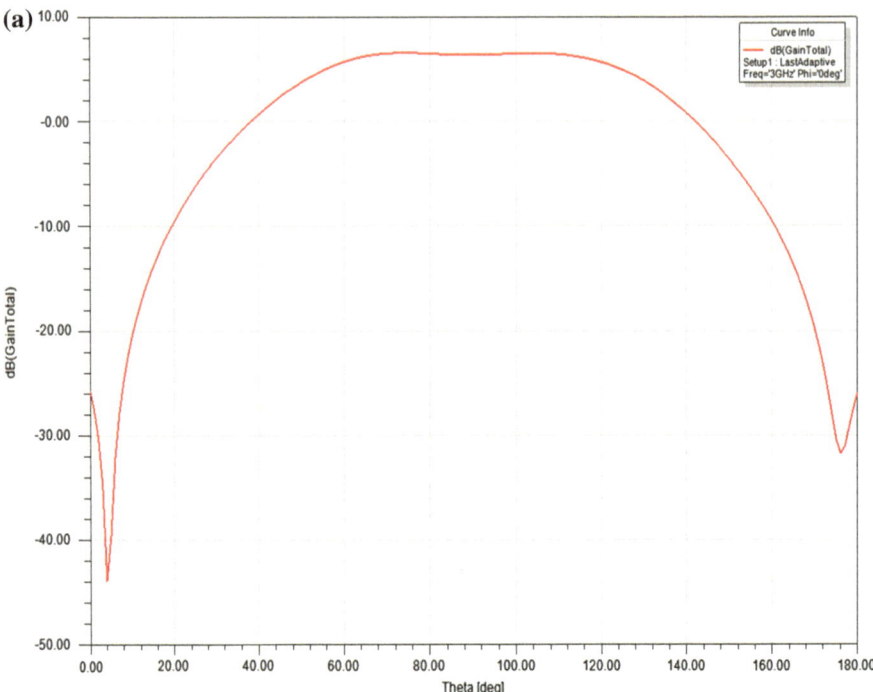

Fig. 42 Radiation pattern of a two-dipole antenna over a cylindrical substrate with truncated ground plane. **a** Rectangular plot. **b** Polar plot. **c** 3-D

4 Planar Dipole Array

Next, the dipole configuration is changed from linear to planar. The dipoles are arranged in two directions (y and z). The design parameters used are same as in Sect. 3.4 except the inter-element spacing along the z-direction.

4.1 A Square Dipole Array Over a Cylindrical Surface

A square array (3×3) is designed using full-wave simulation package. Figure 63 shows the design model of 3×3 dipole array over a cylindrical surface with hollow cylindrical ground plane. The inter-element spacing between the dipoles along z-direction is taken as 0.3λ. One should note that along z-direction, ϕ remains constant. Hence the inter-element spacing will be 0.3λ for each element along z-axis. However, along ϕ-direction the positions of dipole elements vary, given in Table 5.

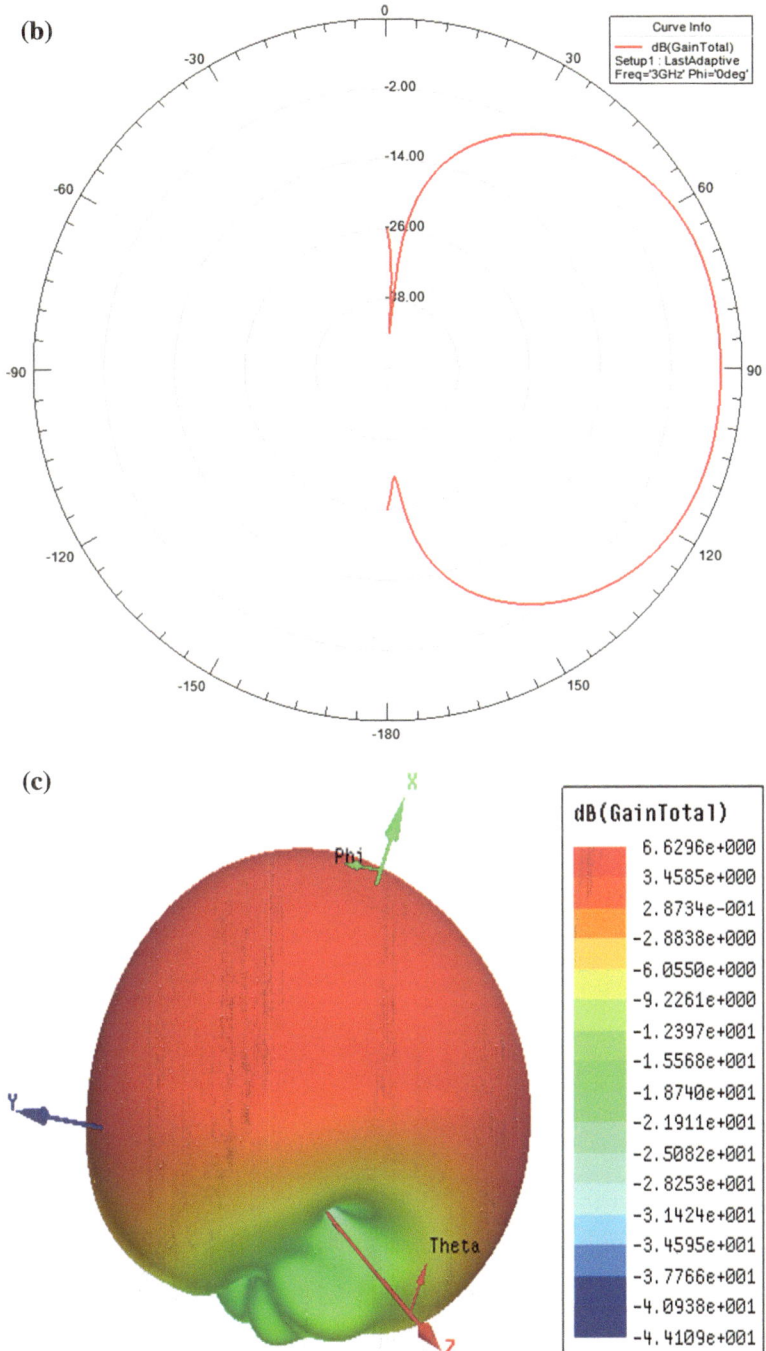

Fig. 42 (continued)

(a)

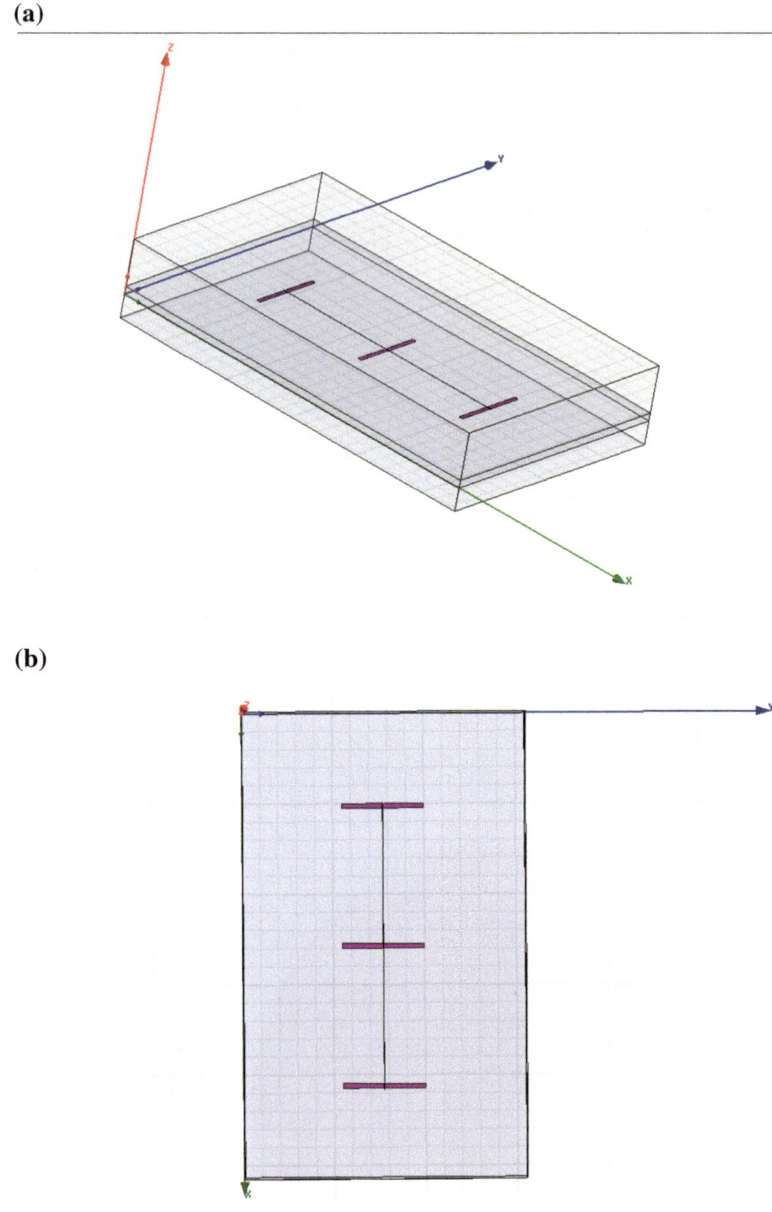

(b)

Fig. 43 Three-dipole array on a planar substrate. a Side view. b Top view

Using the frequency sweep of 1–4 GHz, the variation of real and imaginary parts of antenna impedance of a 3 × 3 squared dipole array over a cylindrical substrate is shown in Fig. 64. The antenna impedance at the input terminals is Re(Z) = 33.73Ω

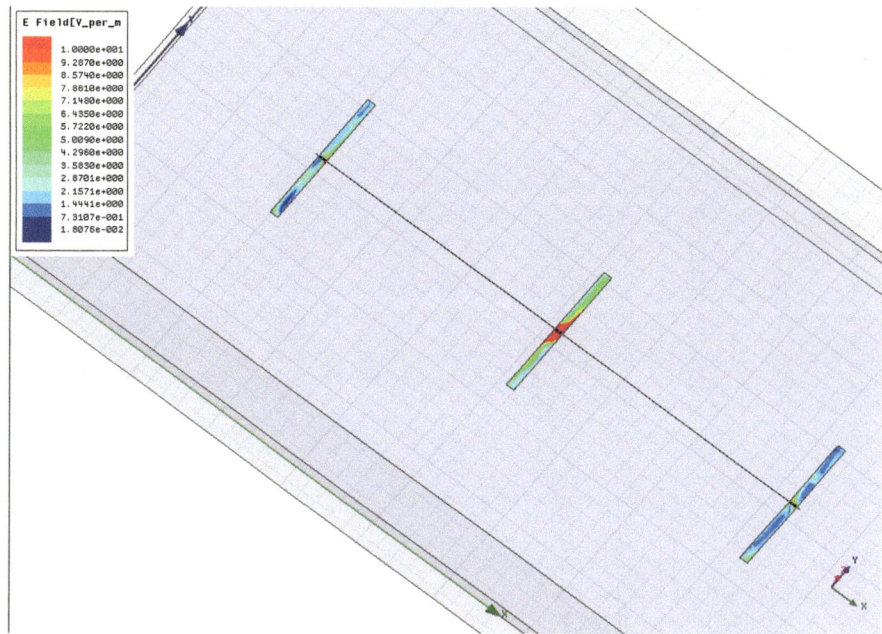

Fig. 44 E-field distribution over three-dipole array antenna

(a)

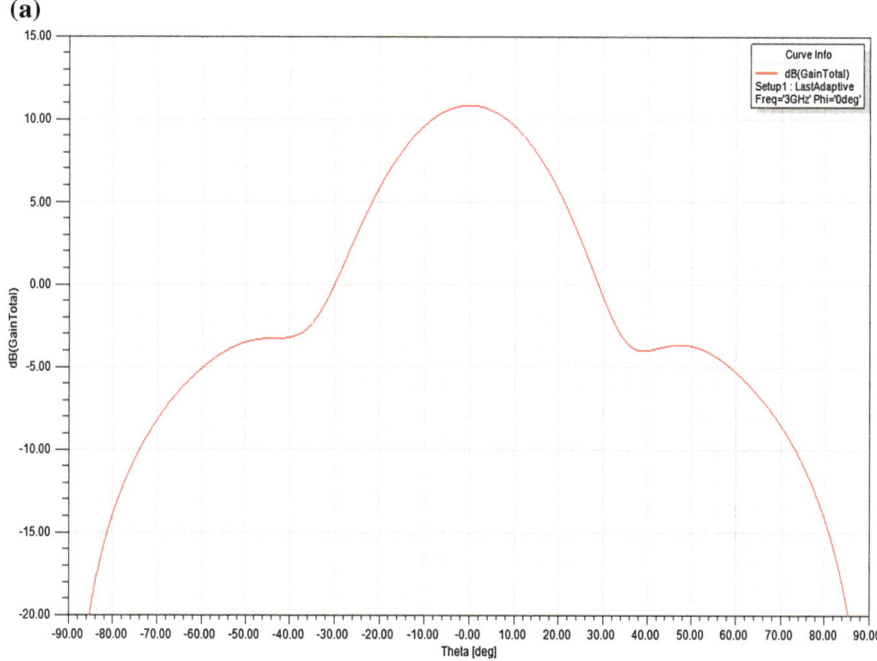

Fig. 45 Radiation pattern of a three-dipole array over a planar substrate and an infinite ground plane. **a** Rectangular plot. **b** Polar plot. **c** 3-D plot

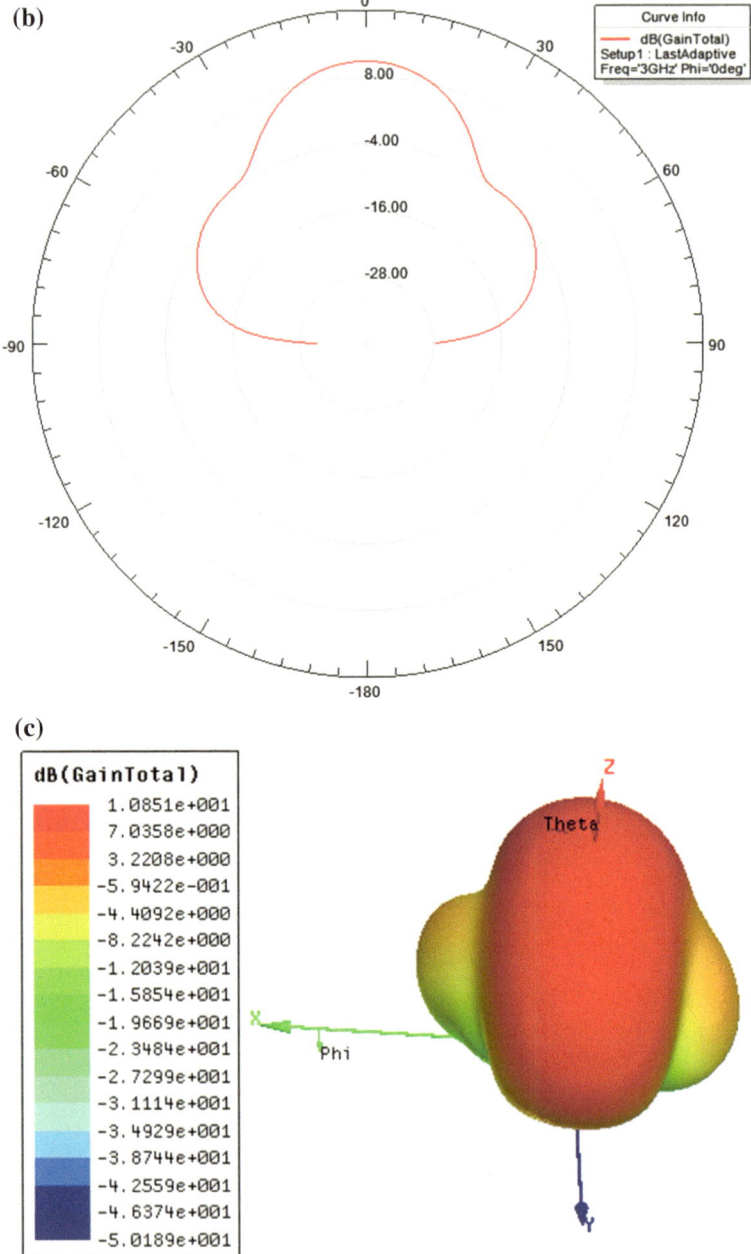

Fig. 45 (continued)

(a)

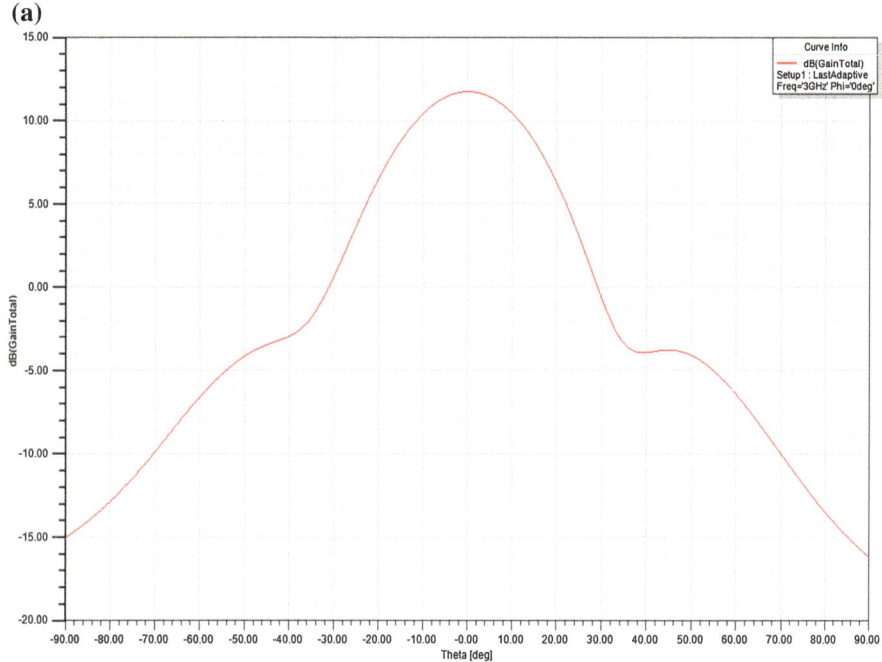

(b)

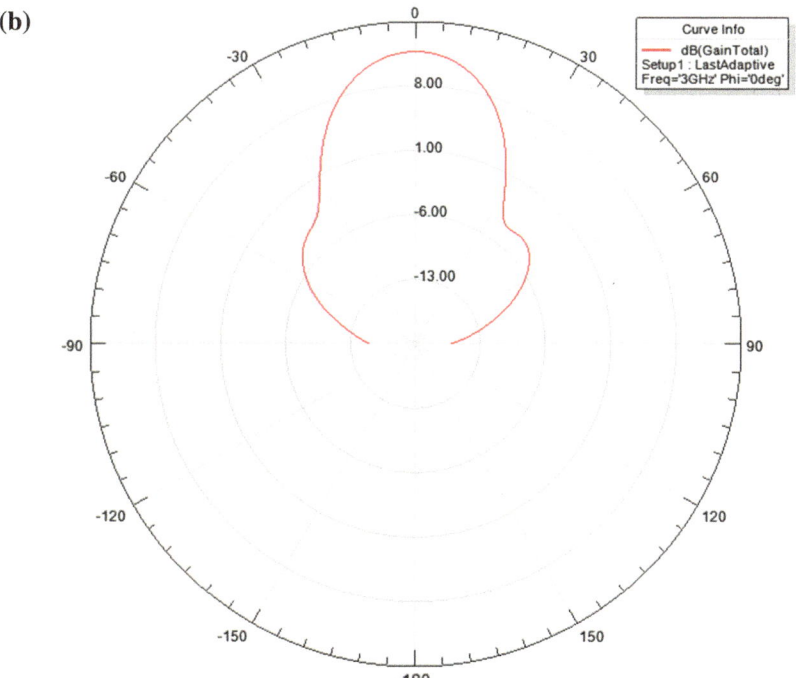

Fig. 46 Radiation pattern of three-dipole array antenna over a planar substrate and finite ground plane. **a** Rectangular plot. **b** Polar plot. **c** 3-D plot

(c)

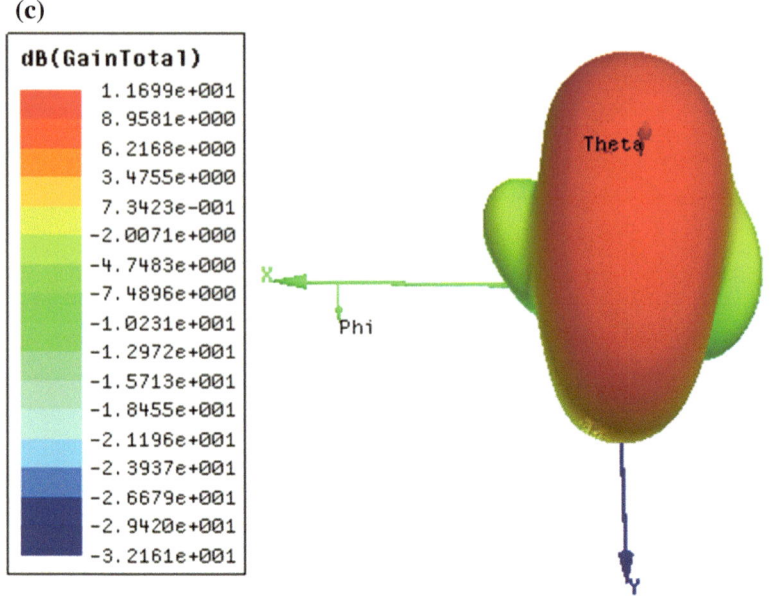

Fig. 46 (continued)

Fig. 47 Three-dipole array over a cylindrical surface with truncated ground plane

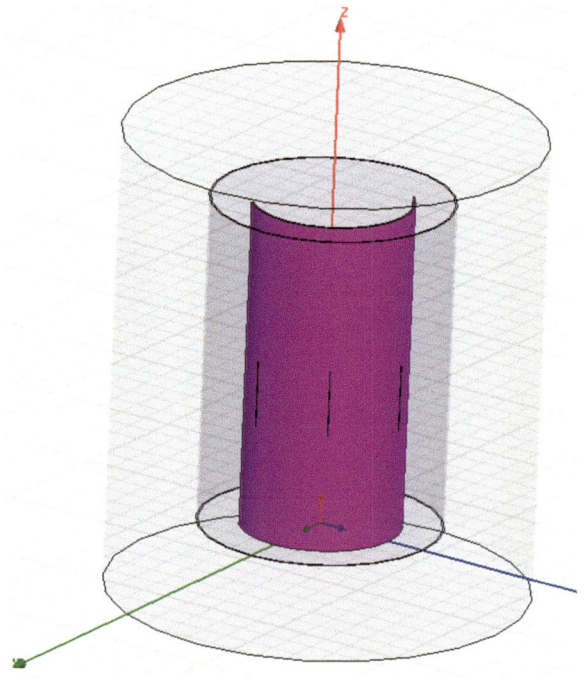

Table 2 Positions of three dipoles on the cylindrical substrate

Arc length, $s = 0.55\lambda$ and radius of cylindrical substrate, $r = 0.9\lambda$

Dipole	ϕ (radians)	Position of dipole on substrate	
		x-coordinate (mm)	y-coordinate (mm)
1	0	90	0
2	0.61	73.768	51.558
3	1.22	30.928	84.518

for Im(Z) = 0. Figure 65 shows the return loss of designed array. It can be observed that the resonant frequency of the dipole array is 2.55 GHz. Figure 66 shows the corresponding radiation pattern of the 3 × 3 dipole array. The radiation pattern is dependent on positions of dipoles.

4.2 A Rectangular Dipole Array Over a Cylindrical Surface

Next, a 2 × 5 planar dipole array is designed on a cylindrical substrate with ground plane. The design parameters are kept same as in Sect. 4.1. Table 6 presents the

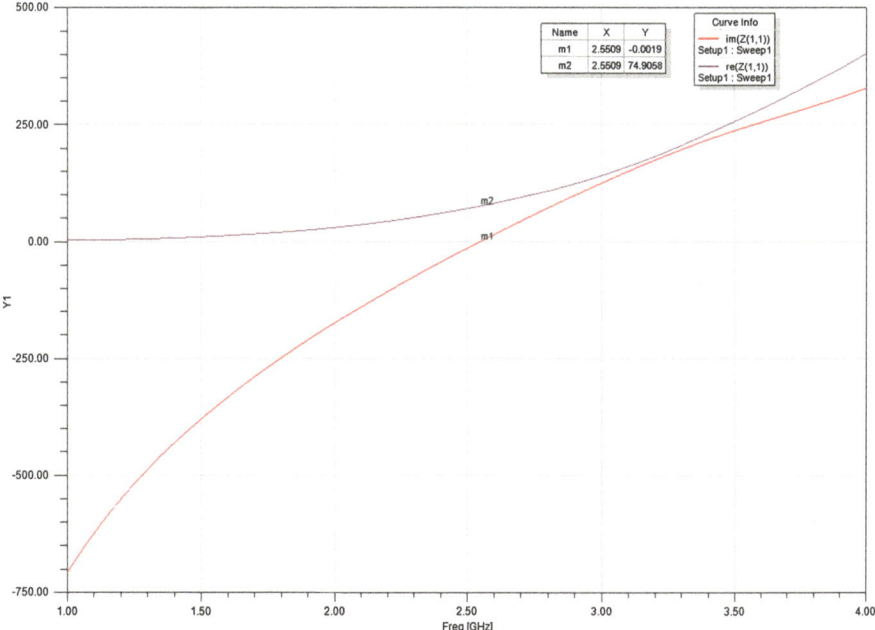

Fig. 48 Impedance (real and imaginary) plot of a three-dipole array over a cylindrical surface with truncated ground plane

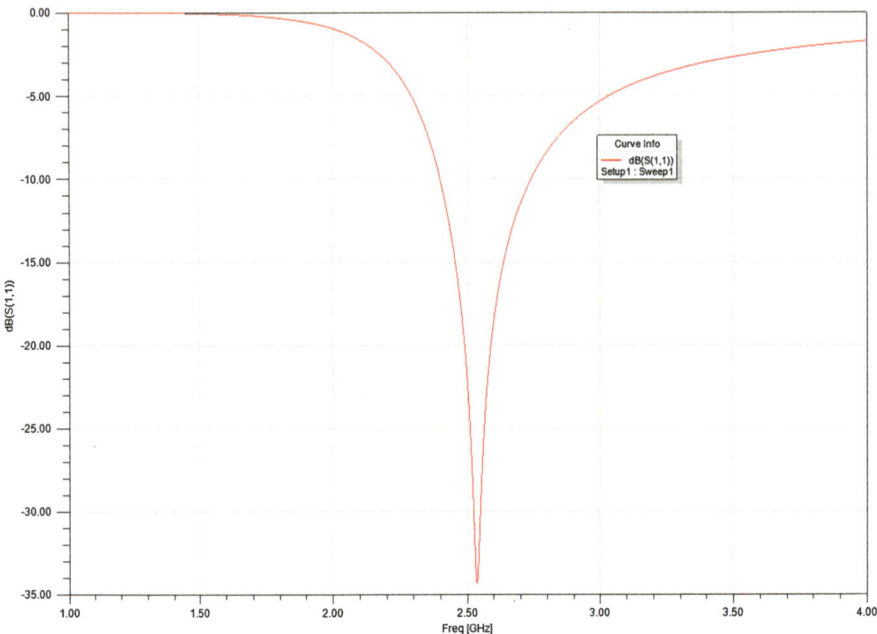

Fig. 49 Return loss of a three-dipole array over a cylindrical surface with truncated ground plane

(a)

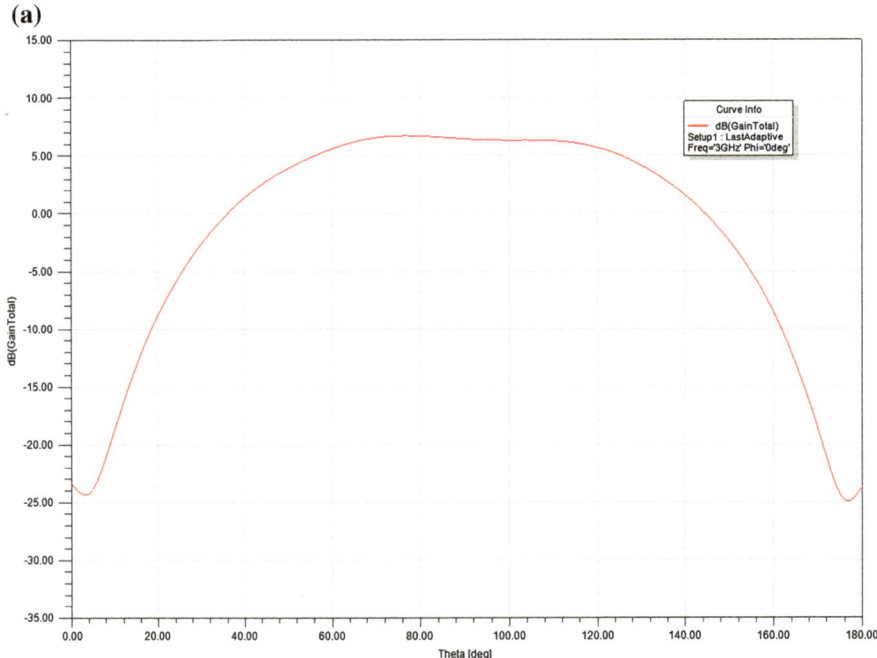

Fig. 50 Radiation pattern of a three-dipole array over a cylindrical surface with truncated ground plane. **a** Rectangular plot. **b** Polar plot. **c** 3-D plot

(b)

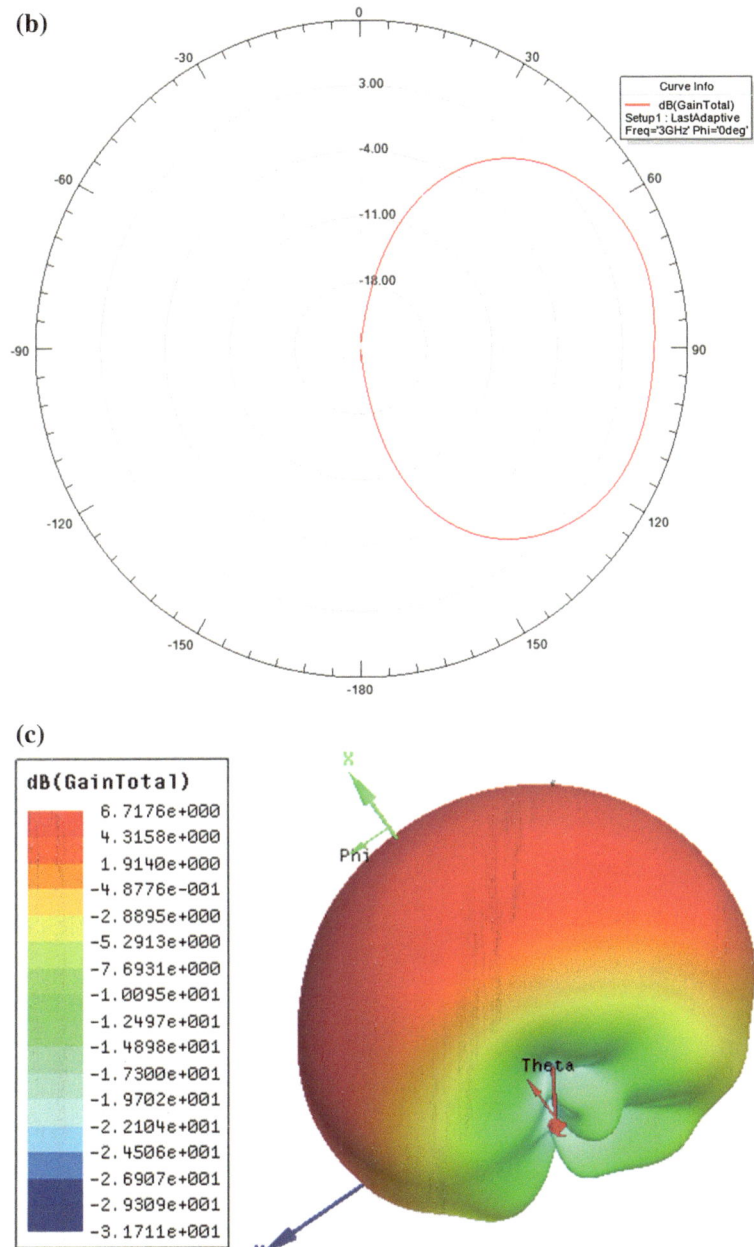

(c)

Fig. 50 (continued)

(a)

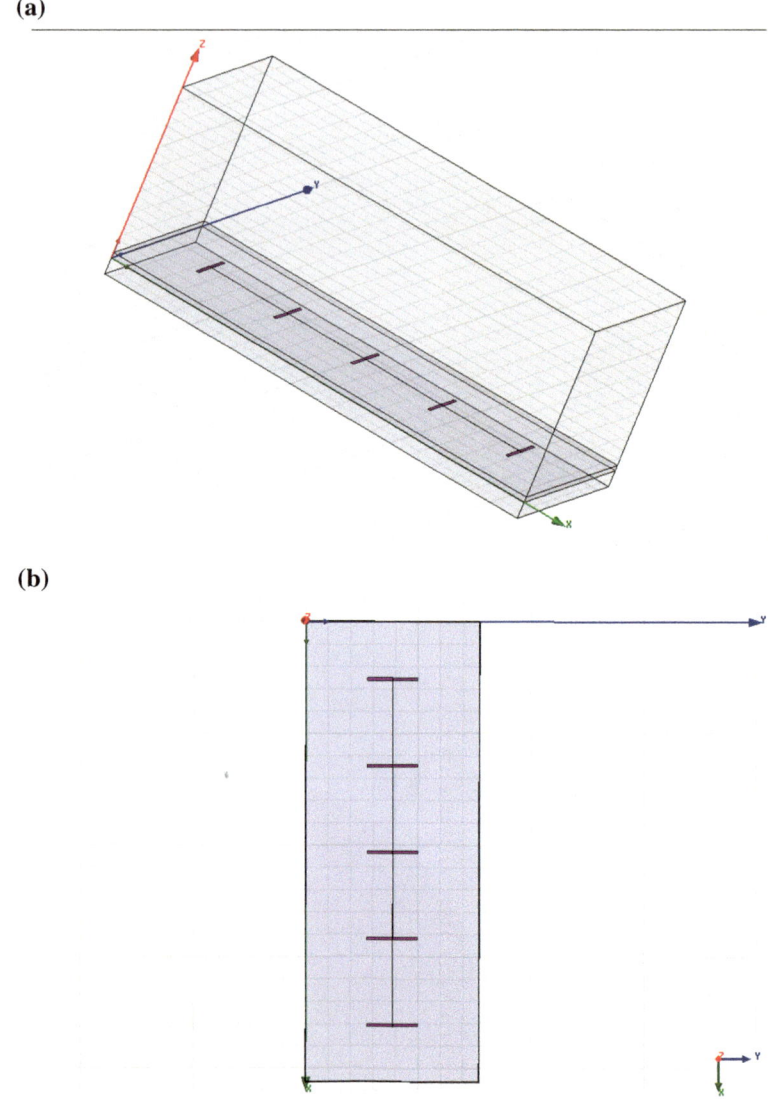

(b)

Fig. 51 Five-dipole array over a planar substrate. **a** Side view. **b** Top view

positions of dipoles on a cylindrical substrate. The design of a 2 × 5 dipole array model is shown in Fig. 67. Using the frequency sweep of 1–3 GHz, the variation of real and imaginary parts of antenna impedance of a 2 × 5 rectangular dipole array over a cylindrical substrate is shown in Fig. 68. The antenna impedance at the input

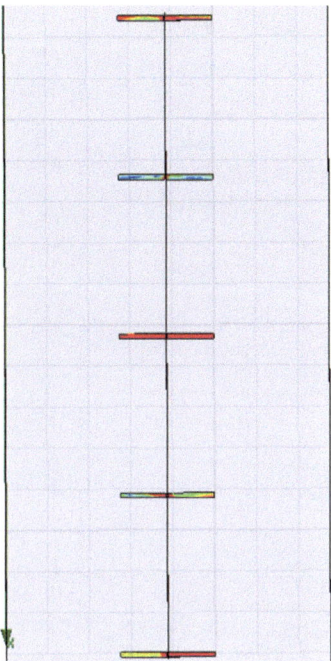

Fig. 52 E-field distribution over a five-dipole array on a planar substrate

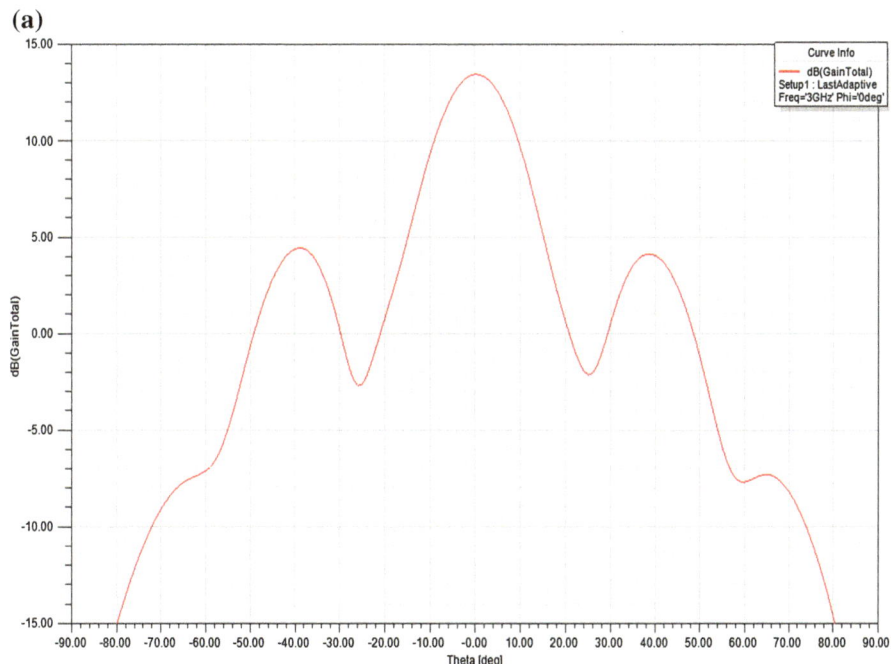

Fig. 53 Radiation pattern of a five-dipole array antenna over a planar substrate and an infinite ground plane. **a** Rectangular plot. **b** Polar plot. **c** 3-D plot

(b)

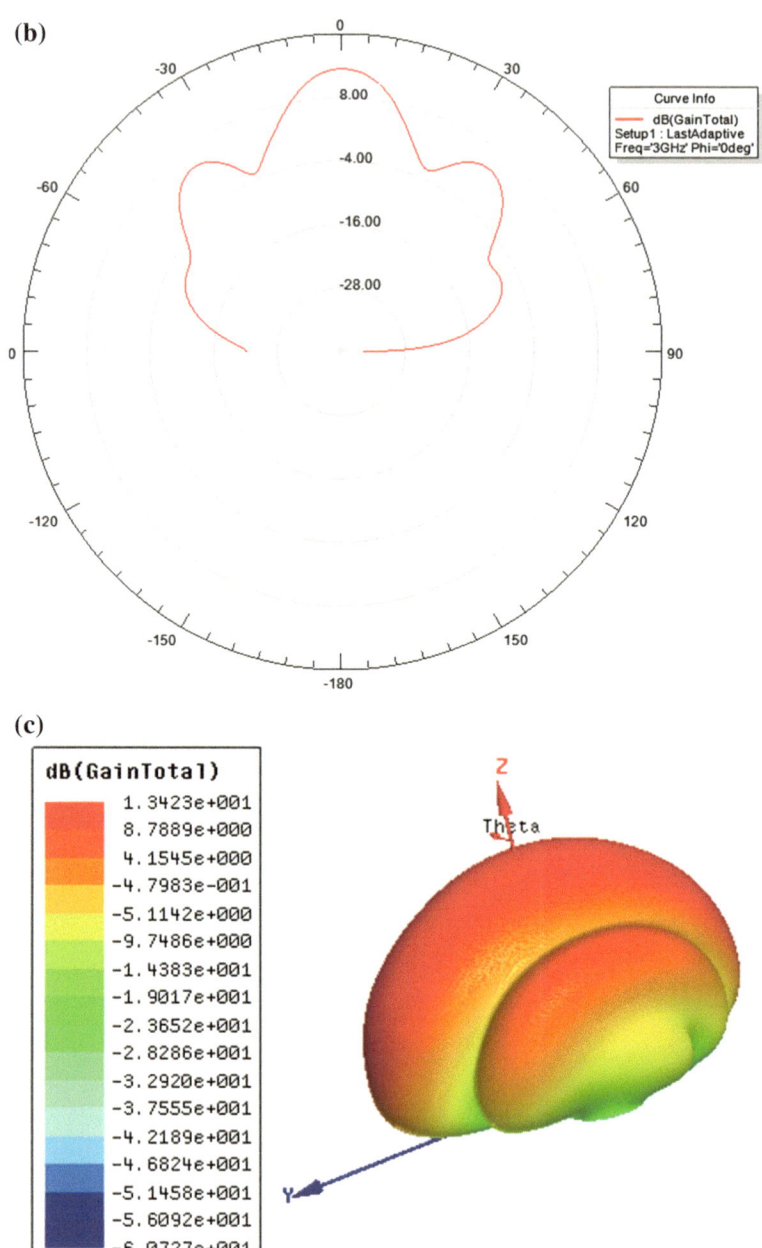

Fig. 53 (continued)

(a)

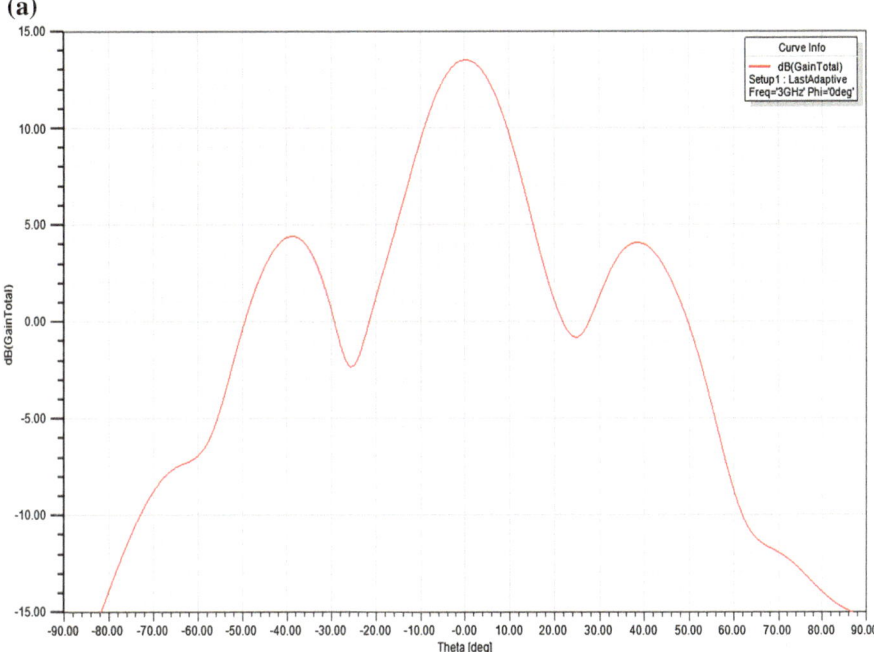

(b)

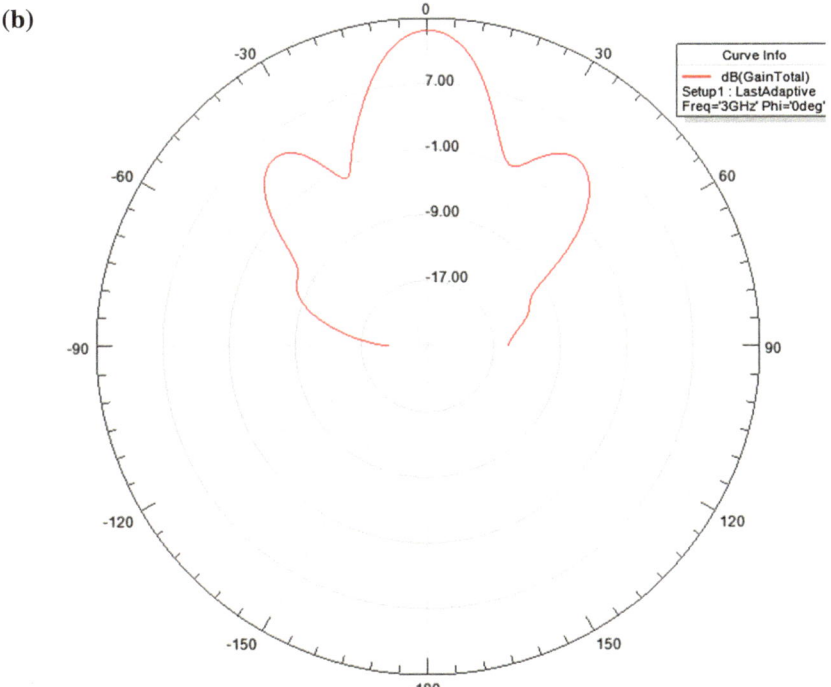

Fig. 54 Radiation pattern of a five-dipole array antenna on planar substrate and finite ground plane. **a** Rectangular plot. **b** Polar plot. **c** 3-D plot

(c)

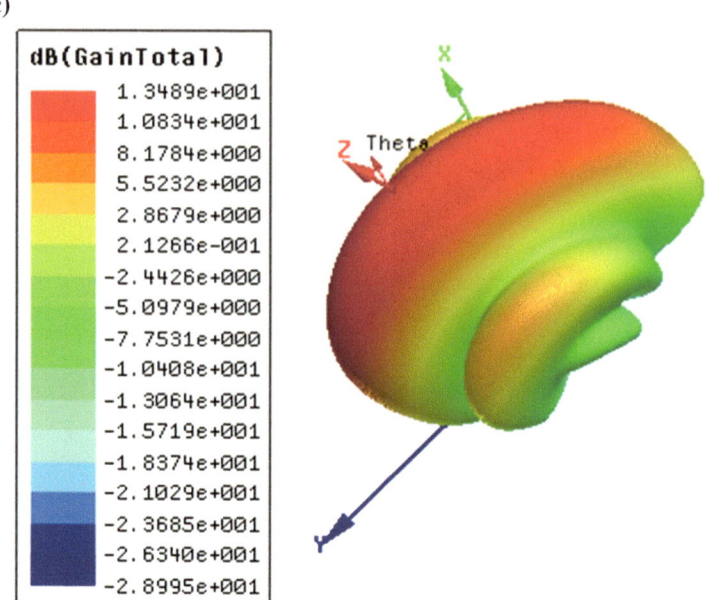

Fig. 54 (continued)

Fig. 55 Five-dipole array
over a cylindrical surface with
truncated ground plane

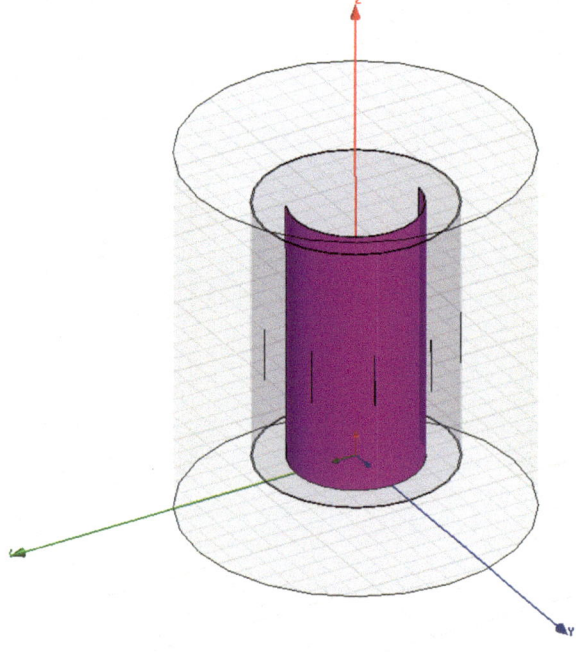

Table 3 Positions of five dipoles on the cylindrical substrate

Arc length, $s = 0.55\lambda$ and radius of cylindrical substrate, $r = 0.9\lambda$

Dipole	ϕ (radians)	Position of dipole on substrate	
		x-coordinate (mm)	y-coordinate (mm)
1	0	90	0
2	0.61	73.768	51.558
3	1.22	30.928	84.518
4	1.83	−23.06	86.993
5	2.44	−68.743	58.089

terminals is $Re(Z) = 31.8849\Omega$ for $Im(Z) = 0$. From the return loss (Fig. 69), the resonance frequency of the dipole array is obtained as 2.55 GHz. The radiation pattern of a 2×5 rectangle dipole array on a cylindrical substrate is shown in Fig. 70.

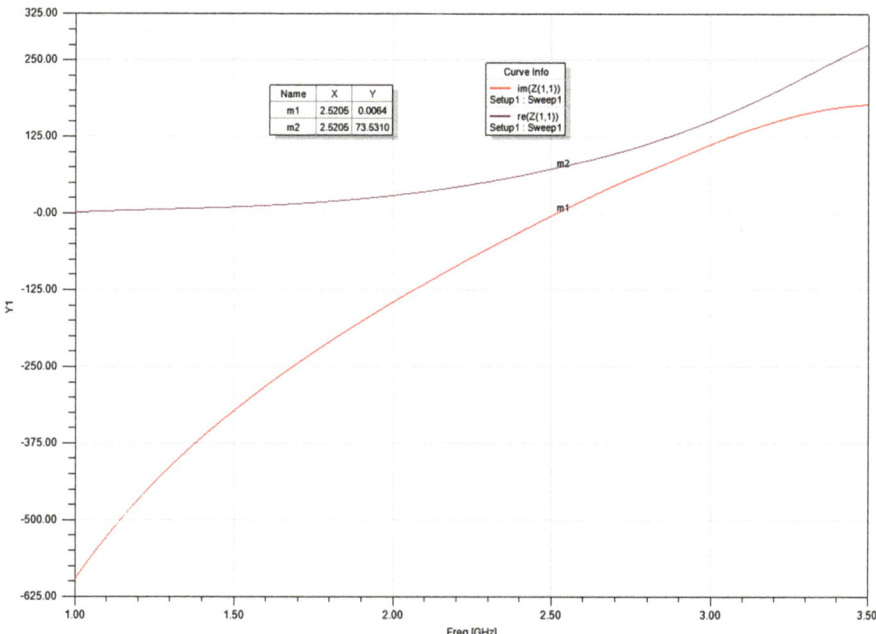

Fig. 56 Impedance (real and imaginary) plot of a five-dipole array over a cylindrical surface with truncated ground plane

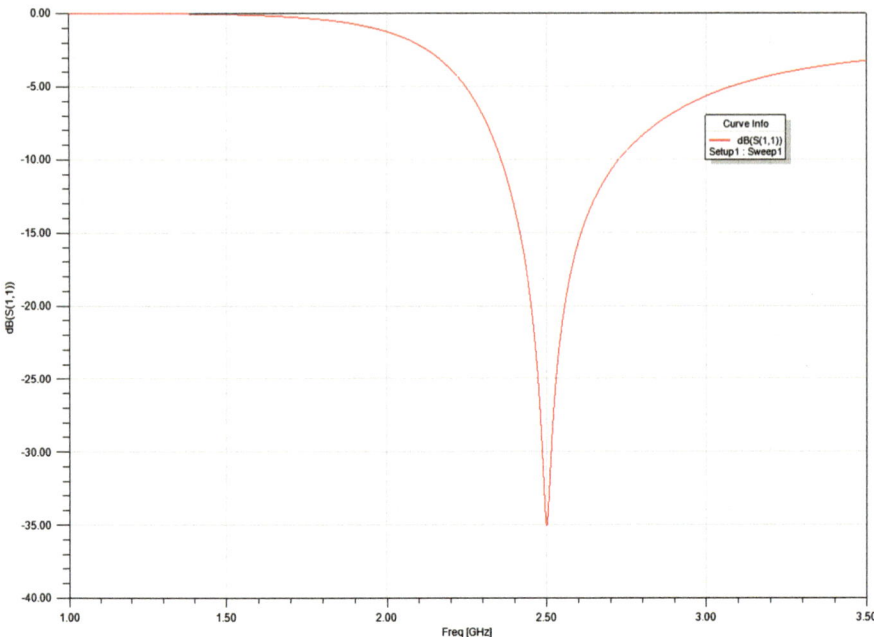

Fig. 57 Return loss of a five-dipole array over a cylindrical surface with truncated ground plane

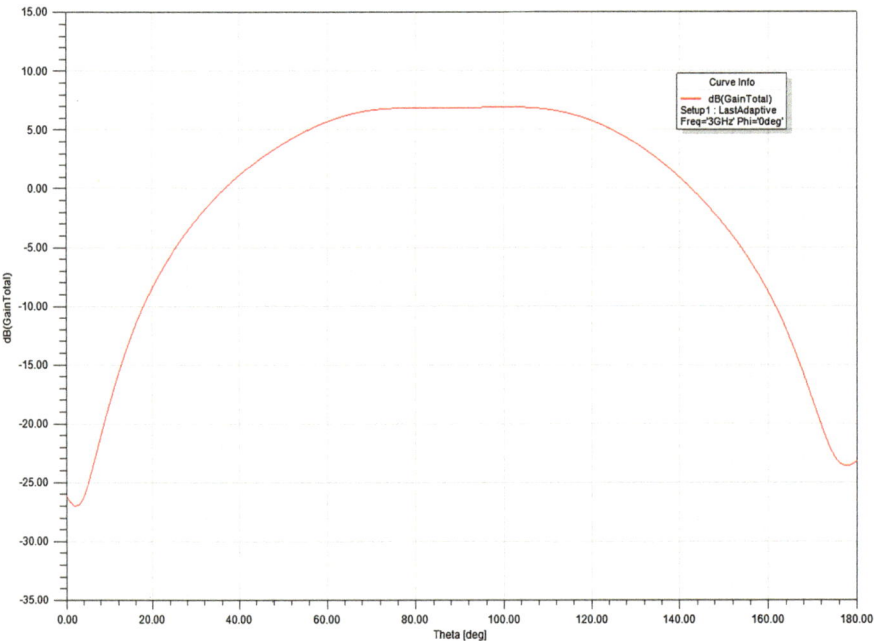

Fig. 58 Radiation pattern of a five-dipole array over a cylindrical surface with truncated ground plane. **a** Rectangular plot. **b** Polar plot. **c** 3-D plot

(b)

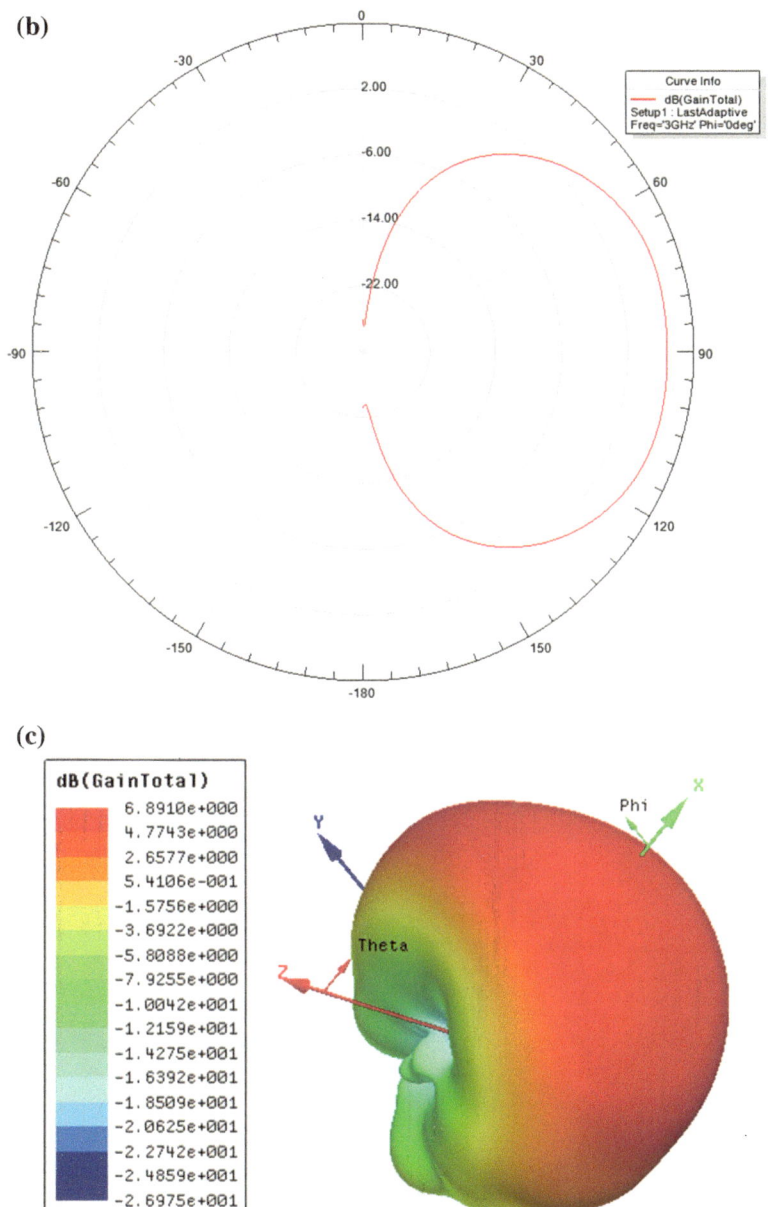

(c)

Fig. 58 (continued)

(a)

(b)

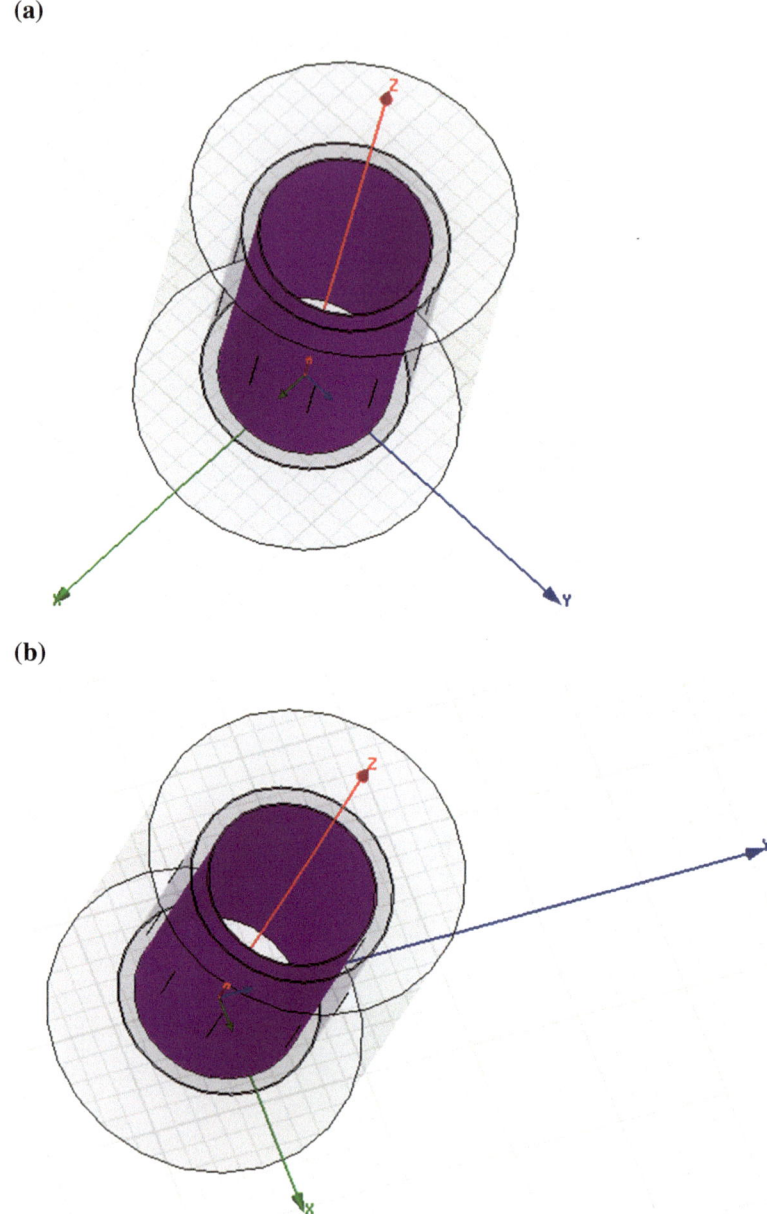

Fig. 59 Ten-dipole array over a cylindrical surface with ground plane. **a** Front view. **b** Back view. **c** Zoomed view

(c)

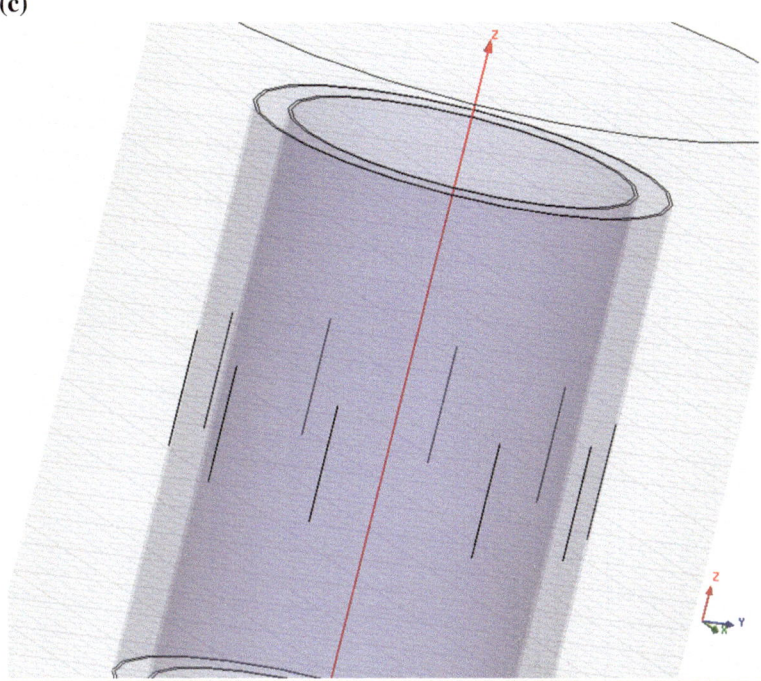

Fig. 59 (continued)

Table 4 Positions of ten dipoles on the cylindrical substrate

Arc length, $s = 0.55\lambda$ and radius of cylindrical substrate, $r = 0.9\lambda$

Dipole	ϕ (radians)	Position of dipole on substrate	
		x-coordinate (mm)	y-coordinate (mm)
1	0	90	0
2	0.61	73.768	51.558
3	1.22	30.928	84.518
4	1.83	−23.06	86.993
5	2.44	−68.743	58.089
6	3.05	−89.6227	8.2318
7	3.66	−78.1748	−44.5947
8	4.27	−38.5289	−81.3358
9	4.88	15.0144	−88.7387
10	5.49	63.1421	−64.1332

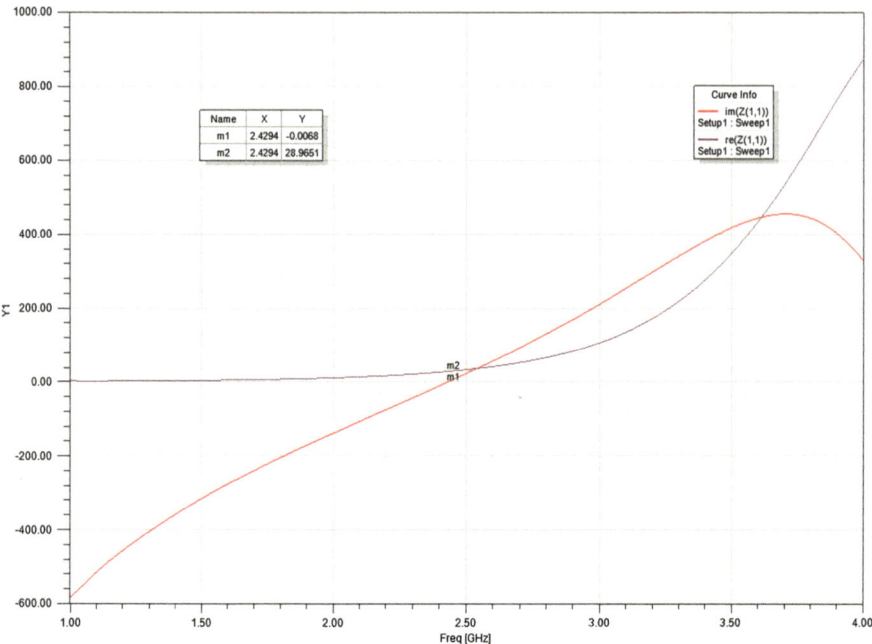

Fig. 60 Impedance (real and imaginary) plot of a ten-dipole array over a cylindrical surface with ground plane

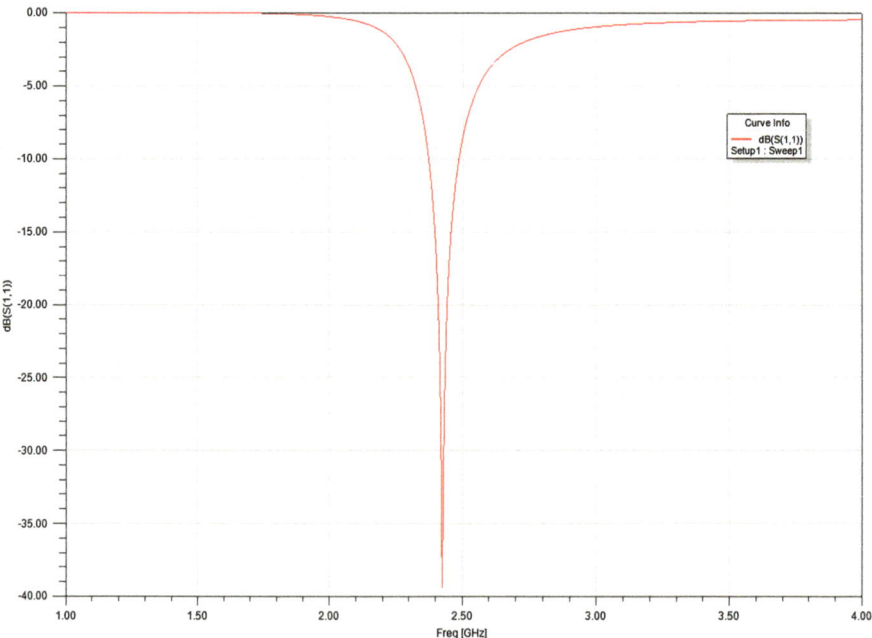

Fig. 61 Return loss of a ten-dipole array over a cylindrical surface with ground plane

(a)

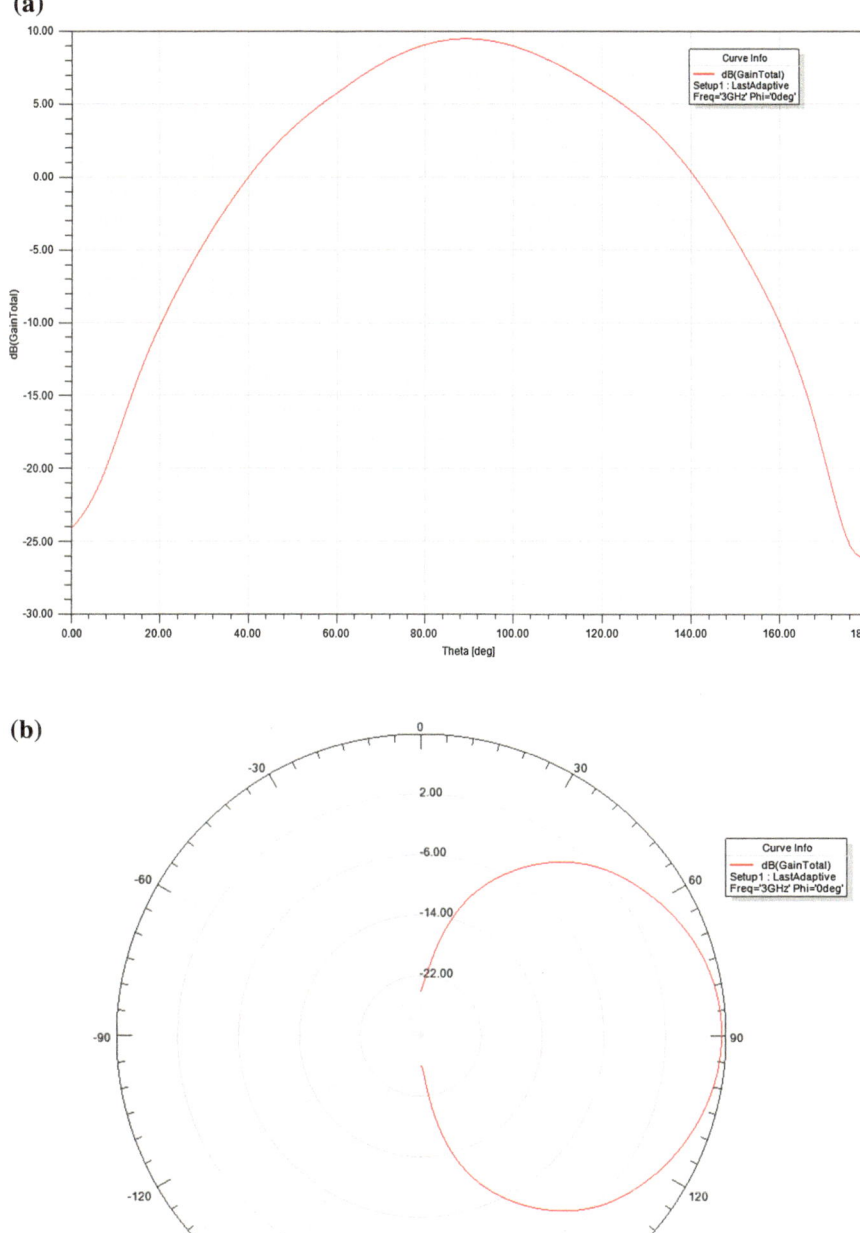

(b)

Fig. 62 Radiation pattern of a ten-dipole array over a cylindrical surface with ground plane

(c)

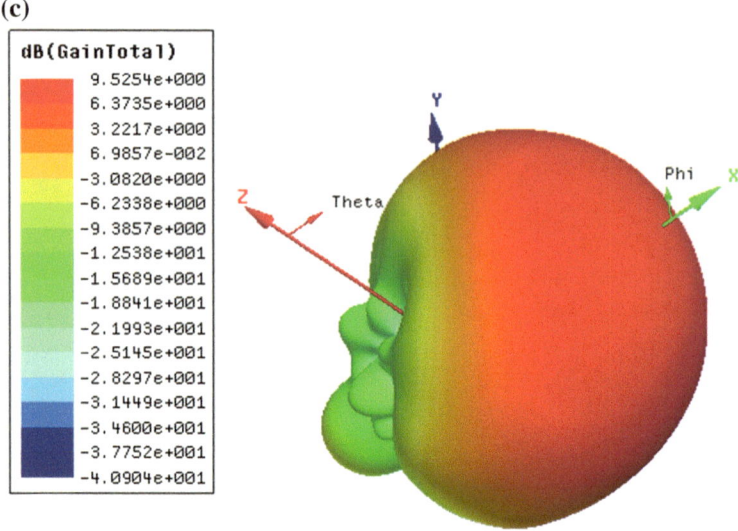

Fig. 62 (continued)

Fig. 63 3 × 3 dipole array over a cylindrical surface with ground plane

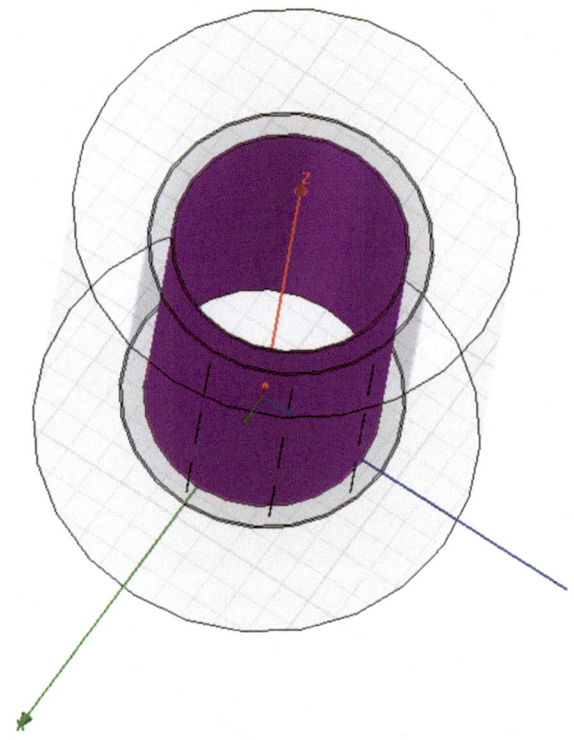

Table 5 Positions of 3 × 3 dipoles on the cylindrical substrate

Arc length, $s = 0.55\lambda$, radius of cylindrical substrate, $r = 0.9\lambda$, inter-element spacing along z-axis = 0.3λ

Dipole		ϕ (radians)	Position of dipole on substrate		
Row	Column		x-coordinate (mm)	y-coordinate (mm)	z-coordinate (mm)
1	1	0	90	0	45.3
1	2	0.61	73.768	51.558	45.3
1	3	1.22	30.928	84.518	45.3
2	1	0	90	0	125.3
2	2	0.61	73.768	51.558	125.3
2	3	1.22	30.928	84.518	125.3
3	1	0	90	0	205.3
3	2	0.61	73.768	51.558	205.3
3	3	1.22	30.928	84.518	205.3

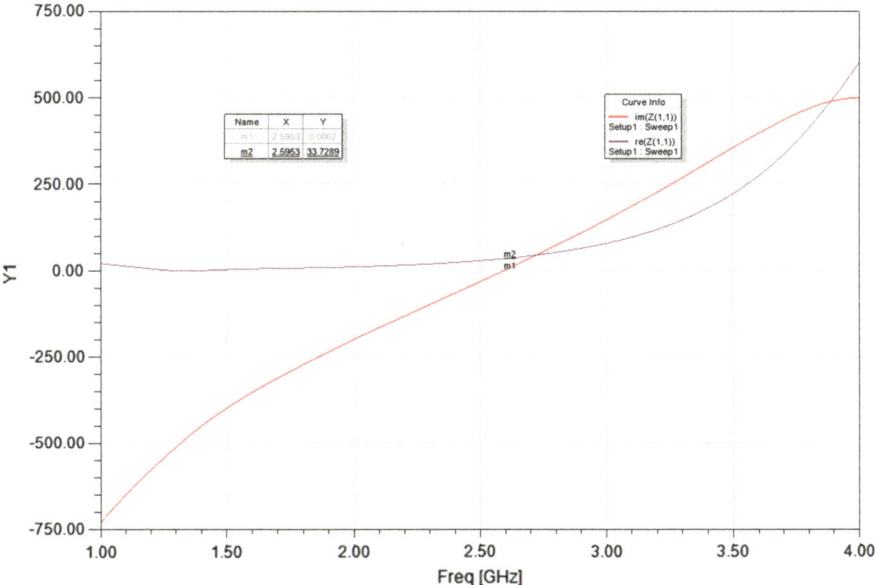

Fig. 64 Impedance (real and imaginary) plot of a 3 × 3 dipole array over a cylindrical surface with ground plane

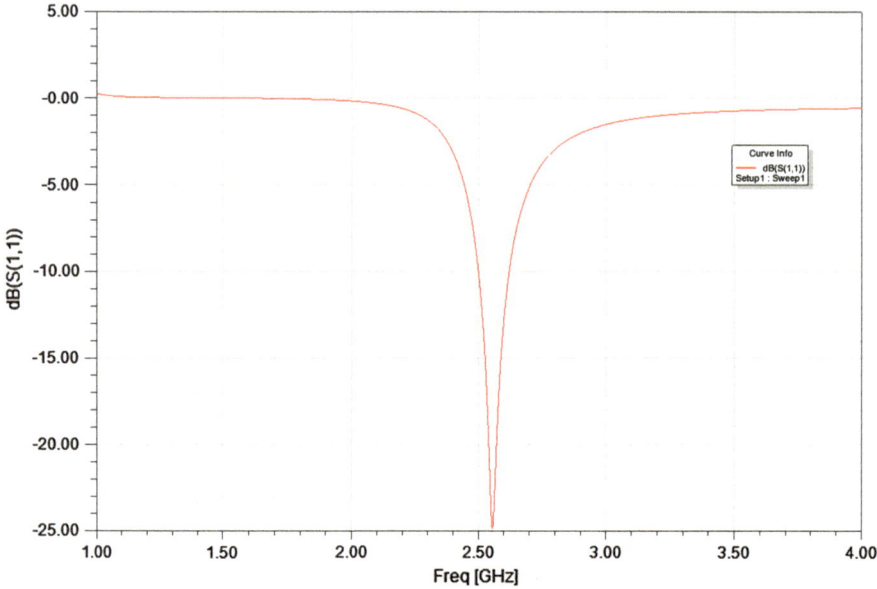

Fig. 65 Return loss of a 3 × 3 dipole array over a cylindrical surface with ground plane

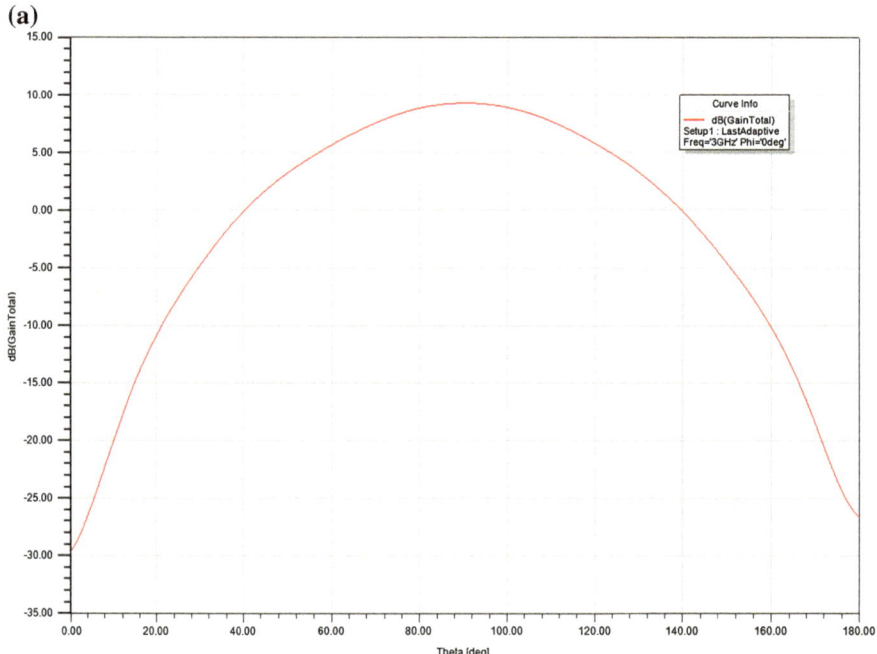

Fig. 66 Radiation pattern of a 3 × 3 dipole array over a cylindrical surface with ground plane. **a** Rectangular. **b** Polar. **c** 3-D

(b)

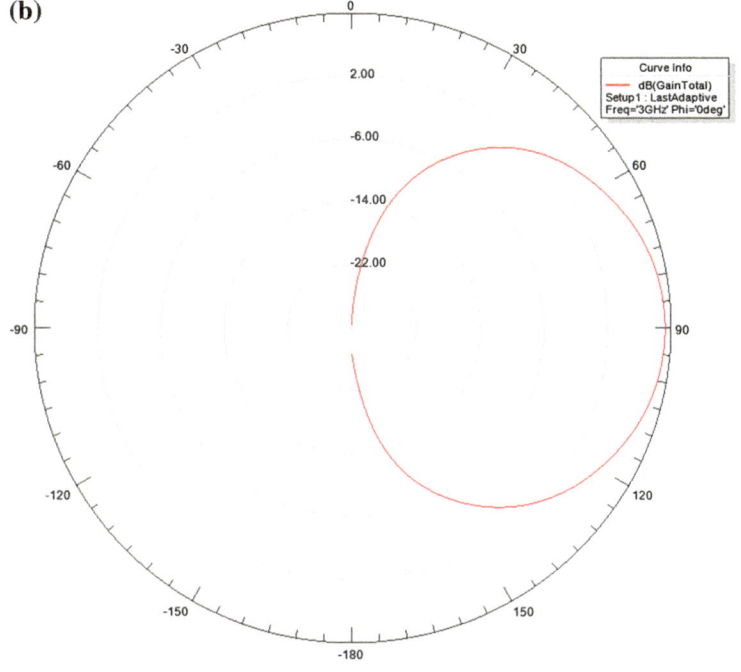

(c)

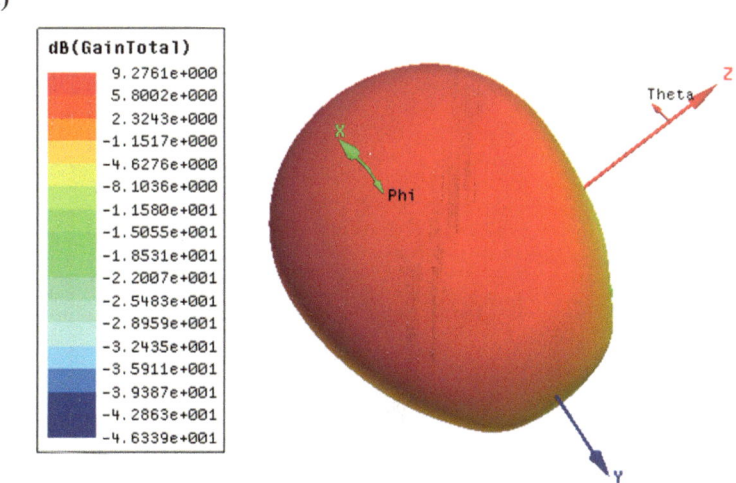

Fig. 66 (continued)

Table 6 Positions of 2 × 5 dipoles on the cylindrical substrate

Arc length, s = 0.55λ, radius of cylindrical substrate, r = 0.9λ, inter-element spacing along z-axis = 0.3λ

Dipole		ϕ (radians)	Position of dipole on substrate		
Row	Column		x-coordinate (mm)	y-coordinate (mm)	z-coordinate (mm)
1	1	0	90	0	125.3
1	2	0.61	73.768	51.558	125.3
1	3	1.22	30.928	84.518	125.3
1	4	1.83	−23.06	86.993	125.3
1	5	2.44	−68.743	58.089	125.3
2	1	0	90	0	205.3
2	2	0.61	73.768	51.558	205.3
2	3	1.22	30.928	84.518	205.3
2	4	1.83	−23.06	86.993	205.3
2	5	2.44	−68.743	58.089	205.3

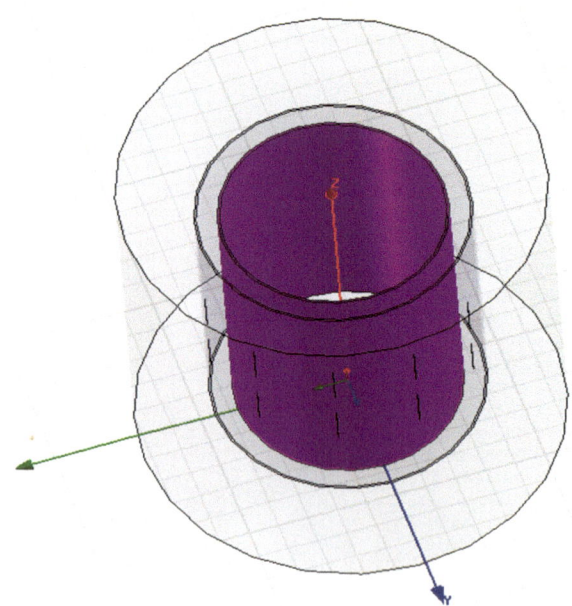

Fig. 67 2 × 5 dipole array over a cylindrical surface with ground plane

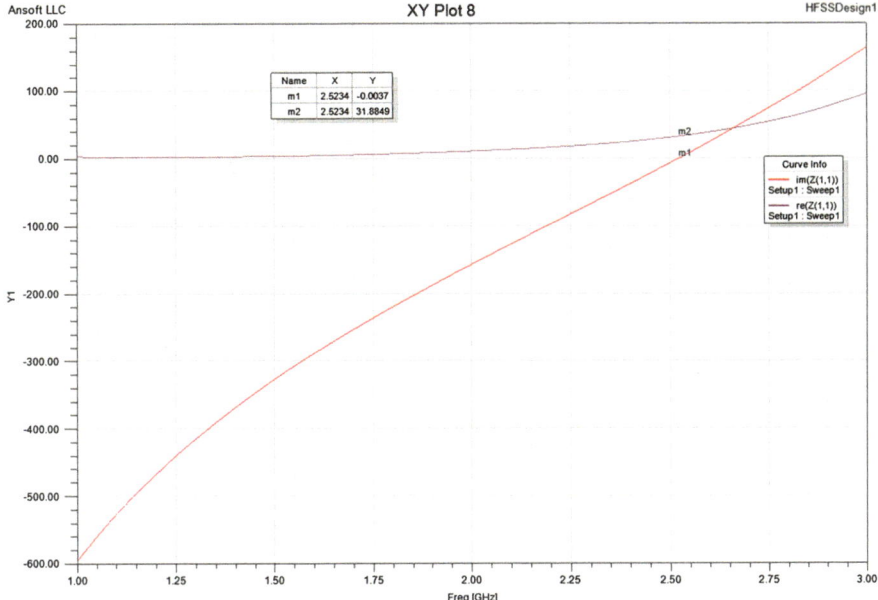

Fig. 68 Impedance (real and imaginary) plot 2 × 5 dipole array over a cylindrical surface with ground plane

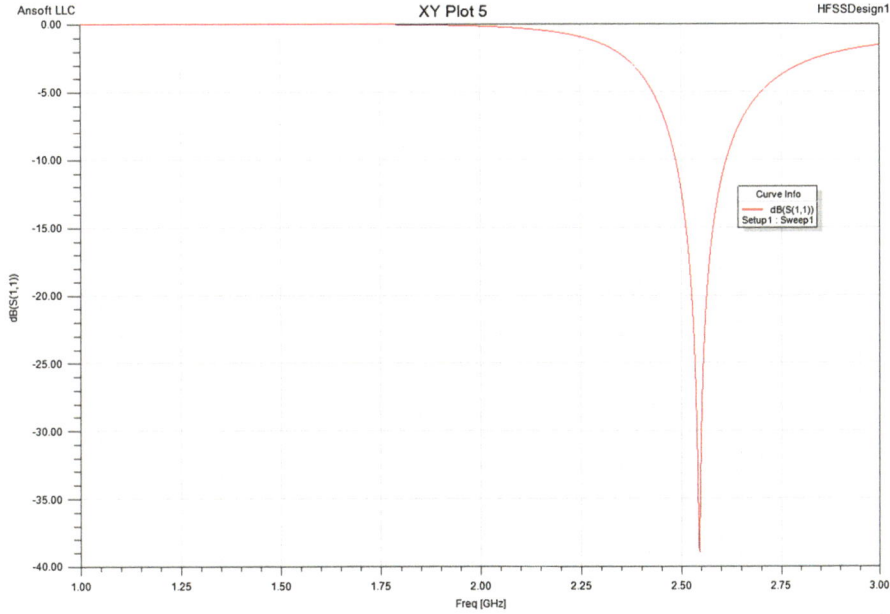

Fig. 69 Return loss of a 2 × 5 dipole array over a cylindrical surface with ground plane

(a)

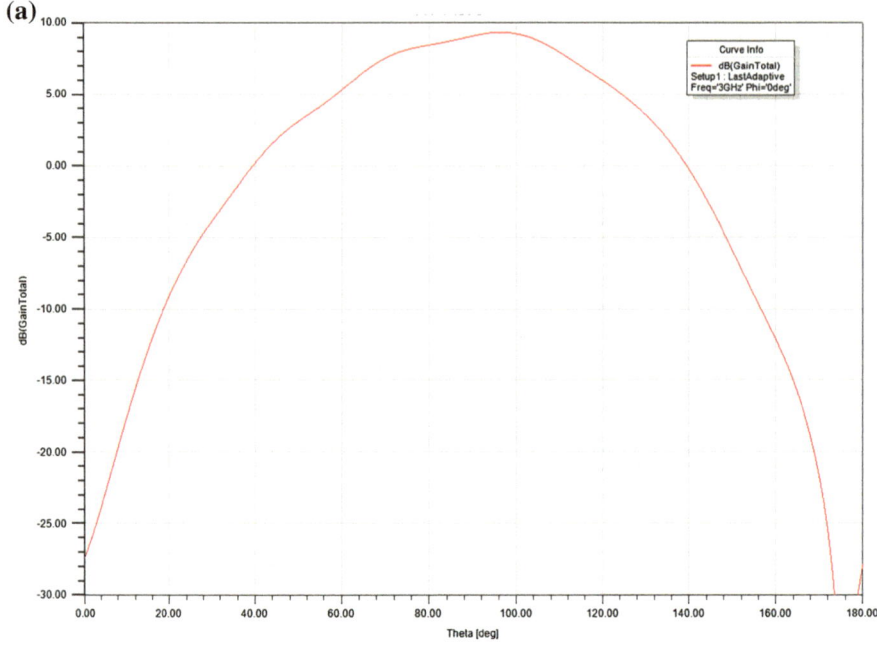

(b)

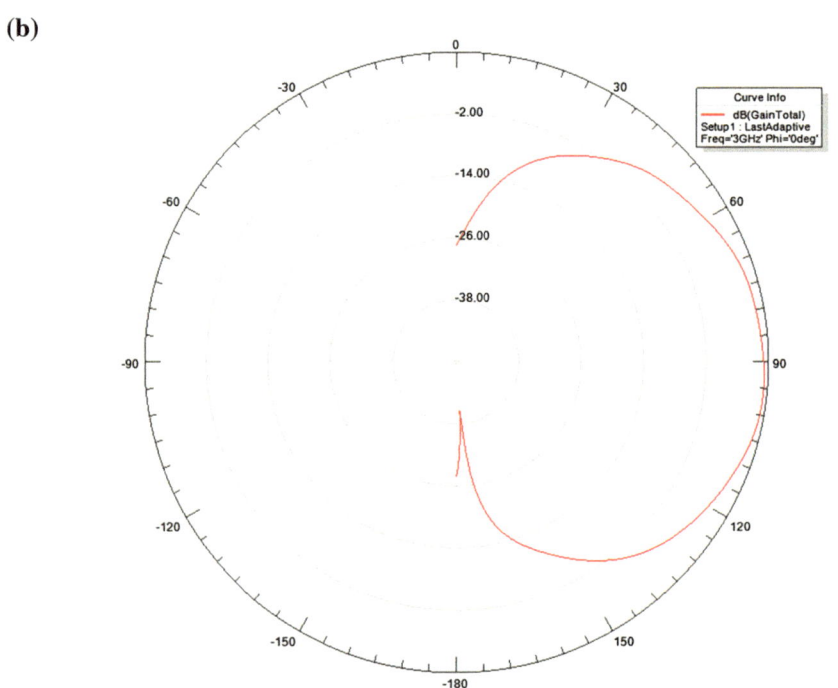

Fig. 70 Radiation pattern of a 2 × 5 dipole array over a cylindrical surface with ground plane. **a** Rectangular. **b** Polar. **c** 3-D

(c)

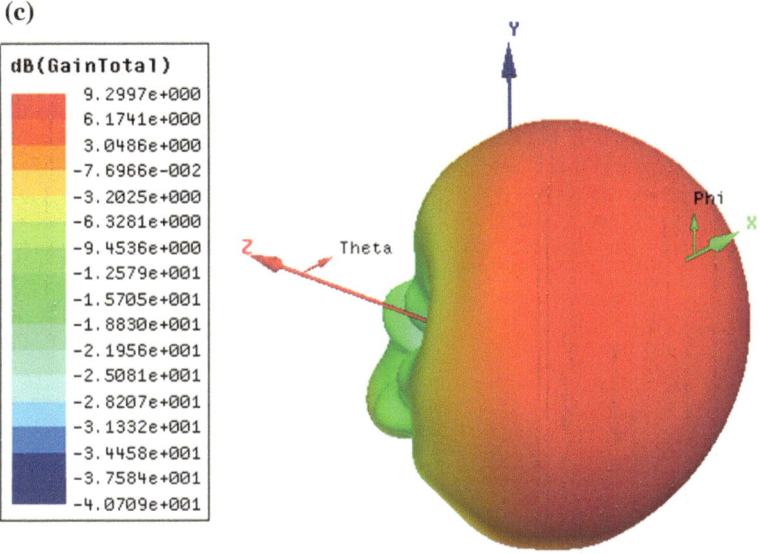

Fig. 70 (continued)

5   Conclusion

Printed dipole antennas are known to be simple but more efficient than the wire antenna. The dielectric substrate and the presence of ground plane affect the antenna performance; the resonant frequency gets shifted. This book presents the EM design of dipole array over planar and cylindrical substrates. The antenna element is taken as half-wave center-fed dipole. The substrate is taken as low-loss dielectric. The design and performance analysis are done using the full-wave simulations. The EM analysis is done for both linear and planar dipole arrays. The input impedance, return loss, and the radiation characteristics are studied for different configurations of dipole array. The objective of this study is toward the design and analysis of conformal array. The effect of curved platform (substrate and ground plane) on the radiation behavior of dipole array is studied. The EM study carried out can further be extended for other curved platforms.

References

Daniil, I.E. 1997. Analysis of finite phased arrays on shaped ground planes. Master Thesis, Naval Postgraduate School, Monterey, CA, 102 p., December 1997.

Elliot, R.S. 1981. *Antenna theory and design.* Englewood Cliffs, NJ: Prentice-Hall Inc., 07632. 594 p. ISBN: 0-13-038356-2.

He, Q.Q., and B.-Z. Wang. 2007. Radiation pattern synthesis for a conformal dipole antenna array. *Progress in Electromagnetics Research, PIER* 76: 327–340.

Kim, J., B.M. Lee and Y.J. Yoon. 2007. Wideband printed dipole antenna for multiple wireless services. *Applied Microwave and Wireless Magazine*, 70–76, September 2007.

Kraus, J.D., R.J. Marhefka and A.S. Khan. 2006. *Antennas for all applications*. 3rd ed. New Delhi: Tata McGraw-Hill, 962 p. ISBN-10: 0-07-060185-2.

Wium, E. 2013. Using CC253X or CC254X with dipole PCB antennas. Design Notes DN041, Texas Instruments, 19 p.

Yang, G.-M., O. Obi, and N.X. Sun. 2012. Enhancing ground plane immunity of dipole antennas with spin-spray-deposited lossy ferrite. *Microwave and Optical Technology Letters* 54: 230–233.

About the Book

This book presents a simple and systematic description of EM design of antenna arrays. Printed dipole antennas are known to be simple yet more efficient than wire antennas. The dielectric substrate and the presence of ground plane affect the antenna performance and the resonant frequency is shifted. This book includes the EM design and performance analysis of printed dipole arrays on planar and cylindrical substrates. The antenna element is taken as half-wave center-fed dipole. The substrate is taken as low-loss dielectric. The effect of substrate material, ground plane, and the curvature effect is discussed. Results are presented for both the linear and planar dipole arrays. The performance of dipole array is analyzed in terms of input impedance, return loss, and radiation pattern for different configurations. The effect of curved platform (substrate and ground plane) on the radiation behavior of dipole array is analyzed. The book explains the fundamentals of EM design and analysis of dipole antenna array through numerous illustrations. It is essentially a step-by-step guide for beginners in the field of antenna array design and engineering.

© The Author(s) 2016   69
H. Singh et al., *EM Design and Analysis of Dipole Arrays on Non-planar Dielectric Substrate*, SpringerBriefs in Computational Electromagnetics,
DOI 10.1007/978-981-287-781-9

Author Index

D
Daniil, I.E., 3

E
Elliot, R.S., 1, 2

H
He, Q.Q., 2

K
Khan, A.S., 2
Kim, J., 2
Kraus, J.D., 2

L
Lee, B.M., 2

M
Marhefka, R.J., 2

O
Obi, O., 2

S
Sun, N.X., 2

W
Wang, B.-Z., 2
Wium, E., 2

Y
Yang, G.-M., 2
Yoon, Y.J., 2

© The Author(s) 2016
H. Singh et al., *EM Design and Analysis of Dipole Arrays on Non-planar
Dielectric Substrate*, SpringerBriefs in Computational Electromagnetics,
DOI 10.1007/978-981-287-781-9

Subject Index

A
Angular coverage, 2
Antenna efficiency, 2
Antenna impedance, 6, 10, 13, 19, 26, 36, 37, 40, 48

B
Boundary condition, 5, 7, 24

C
Constitutive parameters, 2, 3, 7, 21
  loss tangent, 3, 10
  permittivity, 2, 7
Current distribution, 2, 3, 10, 19

D
Dielectric substrate, 1–3, 7, 10, 36, 67
  cylindrical, 2, 3, 9, 10, 18, 19, 21, 26, 33, 35–38, 40, 45, 67
  non-planar, 1, 3, 17
  planar, 1–3, 6, 17, 24, 26, 33, 38, 45
Dipole array, 1–3, 17, 19, 21, 24, 26, 35–40, 45, 53, 67
  linear, 1, 3, 9, 17, 38, 67
  planar, 1–3, 6, 17, 21, 24, 33, 34, 35, 38, 45, 67

E
EM design, 3, 17, 67
EM wave, 2

F
Far field, 5
Feed gap, 3–5, 10
Field distribution, 5, 10, 24, 34

G
Ground plane, 1–7, 10, 13, 18, 24, 26, 34–36, 38, 45, 67
  finite, 3, 5, 7, 24, 35
  infinite, 5–7, 24, 35

I
Inter-element spacing, 18, 19, 36, 38

L
Lumped port, 4, 10, 18, 19, 24, 26, 34, 36, 37

O
Ohmic loss, 2

P
Perfect electric conductor, 2

R
Radiation boundary, 5, 10, 24, 34
Radiation pattern, 2, 3, 5–7, 10, 13, 19, 24, 26, 35–37, 45, 53
Radius of curvature, 2
Resonant frequency, 2, 6, 7, 10, 36, 45, 67
Return loss, 3, 5–7, 10, 13, 17, 19, 26, 36, 37, 45, 53, 67

S
Source, 4
  Current, 4
  Voltage, 4

W
Waveport, 4

© The Author(s) 2016
H. Singh et al., *EM Design and Analysis of Dipole Arrays on Non-planar Dielectric Substrate*, SpringerBriefs in Computational Electromagnetics,
DOI 10.1007/978-981-287-781-9